Nano and Micromachining

Nano and Micromachining

Edited by
J. Paulo Davim
Mark J. Jackson

First published in Great Britain and the United States in 2009 by ISTE Ltd and John Wiley & Sons, Inc.

Apart from any fair dealing for the purposes of research or private study, or criticism or review, as permitted under the Copyright, Designs and Patents Act 1988, this publication may only be reproduced, stored or transmitted, in any form or by any means, with the prior permission in writing of the publishers, or in the case of reprographic reproduction in accordance with the terms and licenses issued by the CLA. Enquiries concerning reproduction outside these terms should be sent to the publishers at the undermentioned address:

ISTE Ltd
27-37 St George's Road
London SW19 4EU
UK

John Wiley & Sons, Inc.
111 River Street
Hoboken, NJ 07030
USA

www.iste.co.uk

www.wiley.com

© ISTE Ltd, 2009

The rights of J. Paulo Davim and Mark J. Jackson to be identified as the authors of this work have been asserted by them in accordance with the Copyright, Designs and Patents Act 1988.

Library of Congress Cataloging-in-Publication Data

Nano and micromachining / Edited by J. Paulo Davim, Mark J. Jackson.
　p. cm.
　Includes bibliographical references and index.
　ISBN 978-1-84821-103-2
　1. Nanotechnology. 2. Micromachining. I. Davim, J. Paulo. II. Jackson, Mark J.
　T174.7.N33 2008
　620'.5--dc22

2008037127

British Library Cataloguing-in-Publication Data
A CIP record for this book is available from the British Library
ISBN: 978-1-84821-103-2

Printed and bound in Great Britain by CPI Antony Rowe, Chippenham, Wiltshire.

Table of Contents

Preface ix

Chapter 1. Nanoscale Cutting 1
Rüdiger RENTSCH

    1.1. Introduction 1
    1.2. Basic elements of molecular dynamics modeling 3
        1.2.1. Material representation and microstructure 3
        1.2.2. Atomic interaction 4
        1.2.3. System dynamics and numerical description 7
        1.2.4. Boundary conditions 8
    1.3. Design and requirements for state-of-the-art MD cutting
    process simulations 10
    1.4. Capabilities of MD for nanoscale material removal
    process analysis 12
        1.4.1. Analysis of microstructure and deformation 12
        1.4.2. Obtaining cutting forces, stress and temperature 15
    1.5. Advances and recent developments in material removal
    process simulation 18
        1.5.1. Complete 3D surface machining simulation 18
        1.5.2. Consideration of fluids in MD cutting simulation 20
    1.6. Summary and outlook 23
    1.7. References 24

**Chapter 2. Ductile Mode Cutting of Brittle Materials: Mechanism,
Chip Formation and Machined Surfaces** 27
Xiaoping LI

    2.1. Introduction 27
    2.2. The mechanism of ductile mode cutting of brittle materials 29
        2.2.1. Transition of chip formation mode from ductile to brittle 29

2.2.2. MD modeling and simulation of nanoscale ductile mode
cutting of silicon................................. 32
2.2.3. The mechanism of ductile mode chip formation in cutting
of silicon..................................... 32
2.3. The chip formation in cutting of brittle materials 35
2.3.1. Material deformation and crack initiation in the chip
formation zone................................. 35
2.3.2. Stress conditions in the chip formation zone in relation
to ductile-brittle mode of chip formation................. 36
2.4. Machined surfaces in relation to chip formation mode 38
2.5. References 40

Chapter 3. Diamond Tools in Micromachining 45
Waqar AHMED, Mark J. JACKSON and Michael D. WHITFIELD

3.1. Introduction.................................... 45
3.2. Diamond technology 45
3.2.1. Hot Filament CVD (HFCVD)..................... 46
3.3. Preparation of substrate............................. 48
3.3.1. Selection of substrate material 48
3.3.2. Pre-treatment of substrate 49
3.4. Modified HFCVD process........................... 51
3.4.1. Modification of filament assembly.................. 51
3.4.2. Process conditions 52
3.5. Nucleation and diamond growth 53
3.5.1. Nucleation 54
3.5.2. Bias-enhanced nucleation (BEN).................... 55
3.5.3. Influence of temperature......................... 56
3.6. Deposition on complex substrates 58
3.6.1. Diamond deposition on metallic (molybdenum) wire....... 58
3.6.2. Deposition on WC-Co microtools 58
3.6.3. Diamond deposition on tungsten carbide
(WC-Co) microtool 59
3.7. Diamond micromachining 62
3.7.1. Performance of diamond-coated microtool............. 66
3.8. Conclusions..................................... 67
3.9. References 67

**Chapter 4. Conventional Processes: Microturning, Microdrilling
and Micromilling** 71
Wit GRZESIK

4.1. Introduction.................................... 71
4.1.1. Definitions and technological possibilities 71
4.1.2. Main applications of micromachining................ 72
4.2. Microturning 74

4.2.1. Characteristic features and applications. 74
4.2.2. Microturning tools and tooling systems. 75
4.2.3. Machine tools for microturning. 77
4.3. Microdrilling. 79
4.3.1. Characteristic features and applications. 79
4.3.2. Microdrills and tooling systems 80
4.3.3. Machine tools for microdrilling. 83
4.4. Micromilling 85
4.4.1. Characteristic features and applications. 85
4.4.2. Micromills and tooling systems. 87
4.4.3. Machine tools for micromilling. 89
4.5. Product quality in micromachining. 92
4.5.1. Quality challenges in micromachining 92
4.5.2. Burr formation in micromachining operations 92
4.5.3. Surface quality inspection of micromachining products 96
4.6. References 98

Chapter 5. Microgrinding and Ultra-precision Processes 101
Mark J. JACKSON and Michael D. WHITFIELD

5.1. Introduction. 101
5.2. Micro and nanogrinding 104
5.2.1. Nanogrinding apparatus. 105
5.2.2. Nanogrinding procedures 105
5.3. Nanogrinding tools 106
5.3.1. Dissolution modeling. 109
5.3.2. Preparation of nanogrinding wheels 110
5.3.3. Bonding systems 112
5.3.4. Vitrified bonding systems 113
5.4. Conclusions. 121
5.5. References 122

Chapter 6. Non-Conventional Processes: Laser Micromachining 125
Grant M. ROBINSON and Mark J. JACKSON

6.1. Introduction. 125
6.2. Fundamentals of lasers 126
6.2.1. Stimulated emission 126
6.2.2. Types of lasers. 127
6.2.3. Laser optics 128
6.2.4. Beam quality. 129
6.2.5. Laser-material interactions 131
6.3. Laser microfabrication 133
6.3.1. Nanosecond pulse microfabrication 133
6.3.2. Shielding gas. 135
6.3.3. Nozzle designs for laser micromachining. 136

6.3.4. Stages of surface melting 138
6.3.5. Effects of nanosecond pulsed microfabrication 138
6.3.6. Picosecond pulse microfabrication. 143
6.3.7. Femtosecond pulse microfabrication 146
6.3.8. Effects of femtosecond laser machining. 150
6.4. Laser nanofabrication. 151
6.5. Conclusions. 154
6.6. References 154

**Chapter 7. Evaluation of Subsurface Damage in Nano
and Micromachining** 157
Jianmei ZHANG, Jiangang SUN and Zhijian PEI

7.1. Introduction. 157
7.2. Destructive evaluation technologies 158
7.2.1. Cross-sectional microscopy 158
7.2.2. Preferential etching. 159
7.2.3. Angle lapping/angle polishing 159
7.3. Non-destructive evaluation technologies 160
7.3.1. X-ray diffraction 160
7.3.2. Micro-Raman spectroscopy 164
7.3.3. Laser scattering 167
7.4. Acknowledgements 172
7.5. References 172

Chapter 8. Applications of Nano and Micromachining in Industry 175
Jiwang YAN

8.1. Introduction. 175
8.2. Typical machining methods 176
8.2.1. Diamond turning 176
8.2.2. Shaper/planner machining. 178
8.3. Applications in optical manufacturing 179
8.3.1. Aspheric lens 179
8.3.2. Fresnel lens 186
8.3.3. Microstructured components 193
8.4. Semiconductor and electronics related applications 200
8.4.1. Semiconductor wafer production. 200
8.4.2. LSI substrate planarization 202
8.5. Summary 203
8.6. Acknowledgements 204
8.7. References 204

List of Authors 209

Index 211

Preface

At this moment in time, it is difficult to obtain an exact definition of nano and micromachining. Nanomachining is a recent nanotechnology that involves changing the structure of nano-scale materials or molecules. The Institute of Nanotechnology (UK) defines nanotechnology as "science and technology where dimensions and tolerances in the range of 0.1 nanometer (nm) to 100 nm play a critical role". Micromachining (performing various cutting processes or grinding operations on workpiece in micro-scale) covers techniques used, for example, in manufacturing the miniaturized devices and moving parts into which microelectronic circuitry is integrated. Unlike micromachining, where portions of the structure are removed or modified, nanomachining involves only changing the structure of nanoscale materials or molecules.

This book aims to provide the fundamentals and the recent advances in nano and micromachining for modern manufacturing and engineering.

Chapter 1 provides the fundamentals of molecular dynamics for nanoscale cutting. Chapter 2 contains information on ductile mode cutting of brittle materials and generic descriptions of the significant aspects involved – mechanism, chip formation and machined surfaces. Chapter 3 covers diamond tools used in micromachining. Chapters 4 and 5 contain information on convention machining processes, microturning, microdrilling, micromilling, microgrinding and ultra-precision processes. Chapter 6 focuses on a non-conventional process – laser micromachining. Chapter 7 covers the evaluation of subsurface damage in nano and micromachining. Finally, Chapter 8 is dedicated to applications of nano and micromachining in industry.

The present book can be used as a textbook for a final year undergraduate engineering course or specifically for nano and micromanufacturing (machining) at postgraduate level. Also, this book can serve as a useful reference for academics, manufacturing and materials researchers, manufacturing and mechanical engineers, as well as professionals in nano and micromanufacturing and related industries. The scientific interest of this book is evident for many important research centers, laboratories and universities in the world. Therefore, it is hoped that this book will encourage and enthuse other research in this recent field of science and technology.

The editors acknowledge their gratitude to ISTE-Wiley for this opportunity and for their professional support. Finally, we would like to thank all the chapter authors for their availability for this work.

J. Paulo Davim
University of Aveiro, Portugal
October 2008

Mark J. Jackson
Purdue University, USA
October 2008

Chapter 1

Nanoscale Cutting

1.1. Introduction

In nano and micromachining processes the actual material removal can be limited to the surface of the workpiece, i.e. only a few atoms or layers of atoms. At this range, inherent measurement problems and the lack of more detailed experimental data limit the possibility for developing analytical and empirical models as more assumptions have to be made. On the basis of atomistic contact models, the dynamics of the local material removal process and its impact on the material structure, as well as the surface generation, can be studied.

The first pioneering applications in molecular dynamics (MD) indentation and material removal simulation were published between 1989 and 1991 [BEL 91, IKA 91, HOO 90, LAN 89]. By starting at the atomic level, the considered microscopic material properties and the underlying constitutive physical equations of state in MD provide, in principle, a sufficiently detailed and consistent description of the micromechanical and thermal state of the modeled material to allow for the investigation of the local tool tip/workpiece contact dynamics [HOO 91, RAP 95]. The description of microscopic material properties considers, e.g., the microstructure, lattice constants and orientation, chemical elements and the atomic interactions.

Chapter written by Rüdiger RENTSCH.

2 Nano and Micromachining

The more universal material representation in MD further allows us to go beyond ideal, single crystalline structures and to also consider polycrystals, defect structures, pre-machined or otherwise constrained workpiece models and non-smooth surfaces [DAW 84, REN 95-1, REN 95-3, YIP 89]. Various application-specific boundary conditions may be applied [HOO 91, RAP 95, YIP 89]. In recent years the number of applications considering quantum mechanics for the interactions between atoms has been steadily increasing. However, here only the more classical atomistic approach will be presented.

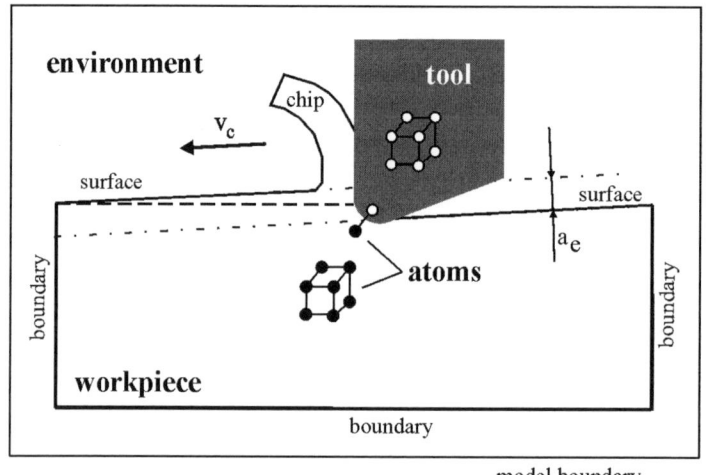

Figure 1.1. *Concept of a molecular dynamics cutting model setup*

Figure 1.1 shows a general description of an often applied concept for MD cutting process simulation, i.e. the orthogonal cutting condition, and includes the essential elements of MD modeling. In addition to the material properties and the interactions between its constituents, the contact and interface conditions, e.g. between tool tip and workpiece as well as with their environment, need to be described. Furthermore, the boundary conditions within the model (surfaces vs. bulk material) and the system boundaries to the non-modeled environment are of importance. Table 1.1 provides a list of the necessary physical elements and principles as well as their area of application in MD modeling. The mathematical description of the equation of motion in particular has been included in this listing, since its choice has a major influence on the numerical complexity and the accuracy of calculation.

In respect of the application of MD modeling for the nanoscale cutting process simulation, in the following chapter some of the basic elements in Table 1.1 will be described in more detail first. Then, in section 1.3, the design and requirements for state-of-the-art MD cutting process simulations will be discussed and, in the following section, the capabilities of MD for the nanoscale material removal process analysis will be demonstrated on the basis of results of application examples. Some aspects regarding significant advances and recent developments in MD material removal process simulation will be discussed in section 1.5, before the summary and outlook of this contribution is given.

Physical element/ principle	Application in MD
• microstructure	initial configuration
• micro-mechanics	atomic interaction
• dynamics	equation of motion
• mathematical description	numerical integration (dynamics)
• thermodynamics	energy balance of the system
• boundary conditions	micromechanical boundaries of the model

Table 1.1. *Application area of the physical elements and principles in MD modeling*

1.2. Basic elements of molecular dynamics modeling

1.2.1. *Material representation and microstructure*

While the original molecular dynamics theory is well based within physics, empirical elements were introduced from the materials science field in order to match the results of experiments with the theoretical and so far physical model. The key to computational efficiency of atomic-level simulations lies in the description of the interactions between the atoms at the atomistic instead of the electronic level. This reduces the task of calculating the complex many-body problem of interacting electrons and nuclei as in quantum mechanics to the solution of an energetic relation involving, basically, only atomic coordinates [HOO 91]. Accordingly, a discrete body or a certain material is described by its chemical elements and by their coordinates. The coordinates provide the information about the atomic arrangement, i.e. the structure of the material, which could be set up, e.g. for a metal on the basis of known lattice structures and lattice constants.

The atomic arrangements in Figure 1.1 hint at the requirement of a description for all matter involved, primarily for the workpiece and the tool material. Considering the crystal size of typical metals, which range between a few tens to several hundred microns in diameter, single crystalline workpiece structures represent reasonable material structures for nanoscale cutting simulations as the tool tip will have to cut over a length of at least 30,000 unit cells before reaching a grain boundary area. However, defects in crystalline structures, like grain boundaries and dislocations [DAW 84, REN 95-3, REN 06, SHI 94, YIP 89], amorphous materials [GLO 95, RAP 95] or polymers as well as liquids and gases [ALL 87, RAP 95] can also be studied using MD. Although Figure 1.1 shows a 2D orthogonal cutting setup, the choice of material representation should always be 3D, even if the width of the model is chosen to be only one unit cell wide. The advantage of 2D models lies in the reduced calculation time and a somewhat easier visualization of the results. However, these advantages are combined with many disadvantages and a great loss of information and meaning of carrying out atomistic simulations. With pure 2D models it is impossible to sufficiently describe the 3D crystalline structure of metals and, hence, no realistic slip system or dislocation motion seems possible and no realistic deformation behavior can be expected. Because of the missing third dimension, 2D simulations result in enhanced, deeper deformation slip as atoms are constrained to accommodate within a plane, in opposition to a 3D model, where each atom has an additional degree of freedom (DoF) to store energy in space [REN 01].

1.2.2. *Atomic interaction*

The central element of the MD code is the calculation of the particle-particle interactions. As it is the most time-consuming part in an MD computer program, it determines the whole structure of the program. Efficient algorithms for the calculation of the interaction are important for systems with a large number of atoms (see [ALL 87, RAP 95]).

The interactions between particles are specified by functions that describe the potential energy. Depending on the complexity of a material and the chosen mathematical description respectively, the potential function may consider many parameters. The goal of the potential function development is that the functional description and the material-specific set of parameters lead to a self-organizing, known structure as a function of the state variables. This provides the basis as well as the necessary flexibility for carrying out not only phase and structure calculations, but also cutting process calculations at the nanoscale. Potential functions and sets of parameters have to be specified for all possible combinations of interactions that need to be considered. In the following, the principles of the necessary potential functions will be described using the widely applied so-called pair potential functions. The class of the more complex many-body potentials, which is of more importance for the representation of metals, will however be discussed only briefly.

1.2.2.1. Pair potentials

First, van der Waals described a model of a material which can form liquid and solid condensed phases at low temperatures and high pressures. Such condensed phases require both attractive and repulsive forces between atoms [HOO 91]. Since the simplest possible representation of many-body interactions is a sum of two-body interactions, the so-called pair potentials were the first potential descriptions of this type. A typical course of the functions is shown in Figure 1.2.

The best known pair potential functions are the Lennard-Jones and the Morse potentials (see equation [1.1] and equation [1.2]) for which the potential energy Φ is only a function of the separation or bonding distance $|r|$ between two atoms. The well-depth of the functions are given by parameters ε and D for the minimum potential energy or sublimation energy, while σ and r_o are constants that define the position of the energy minimum. These parameters are derived from fitting to experimental data like lattice constants, thermodynamic properties, defect energies and elastic moduli. The interaction forces can be derived by calculating the derivative of the potential function, for the pair potential functions only with respect to the separation distance $|r|$.

Lennard-Jones: $\quad \Phi_{LJ}(r) = 4 * \varepsilon * [(\sigma/r)^{12} - (\sigma/r)^6]$ [1.1]

Morse: $\quad \Phi_M(r) = D * [e^{-2\alpha(r-ro)} - 2e^{-\alpha(r-ro)}]$ [1.2]

The potentials describe chemically active materials as bonds that can be established or cut at the long-range part. They represent reasonable descriptions for two-body forces to the extent that they account for the repulsion due to overlapping electron clouds at close distance and for attraction at large distances due to dispersion effects. Generally in solids a shielding effect is expected to make interactions beyond the first few neighbors of limited physical interest. Thus, potential functions are commonly truncated at a certain cutoff distance, preferably with a smooth transition to zero (see Figure 1.2), and result in so-called short-range forces. In addition, the long-range Coulomb forces are usually beyond the reach of MD model sizes [HOO 91].

6   Nano and Micromachining

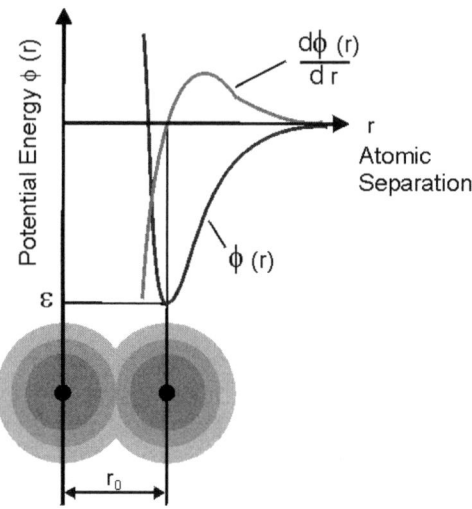

Figure 1.2. *Potential energy according to atomic separation*

1.2.2.2. *Many-body potentials*

The simplicity of the pair potential functions make them appear attractive, for many-atom systems in particular, but they only stabilize structures with equal next neighbor distances, like fcc and hcp structures, basalt planes and triangular lattices. However, using pair-potentials it is not possible to correctly describe all elastic constants of a crystalline metal. For a better representation of metals, many-body interactions need to be included into the function as for example in the well-known potentials following the embedded atom method (EAM) [DAW 84, FIN 84]. In all of the following MD results, the Finnis-Sinclair-type EAM potential by Ackland *et al.* was employed for the workpiece-workpiece interactions [ACK 87]. EAM potentials have been developed and tested for complex problems such as fracture, surface reconstruction, impurities and alloying problems in metallic systems.

The structure of brittle or non-metallic materials with, for instance, covalent or ionic bonds can also not be satisfactorily described by simple pair-potentials. Ionic materials require special treatment because Coulomb interactions have poor convergence properties unless the so-called periodic boundaries are implemented with care (see [ALL 87]). For the diamond lattice or the similar cubic zinc blend structure of covalently bonded semi-conductors like silicon and germanium as well as some ceramics, it is necessary to treat the strong directional bonding explicitly by including terms that describe the interaction between three or more atoms considering bond angles and bond order (see [TER 90, YIP 89]).

1.2.3. *System dynamics and numerical description*

Molecular dynamics comprises macroscopic, irreversible thermodynamics and reversible micro-mechanics. The thermodynamic equations form a link between the micromechanical state, a set of atoms and molecules, and the macroscopic surroundings, the environment. The thermodynamic equations yield the quantities, system temperature and hydrostatic pressure of the model and allow us to determine energy changes involving heat transfer. In mechanics, it is usual to consider energy changes caused by displacement and deformation. By the term "mechanical state" of a microscopic system we mean a list of present coordinates (r) and velocities (v) of the constituents [HOO 91]. For this information about the state of the system to be useful, equations of motion, capable of predicting the future, must be available. As the governing equations of motion for a system of constant total energy, the well-known Newton's equations of motion can be chosen:

Newton's equations of motion $d\{v_i(t)\} / dt = 1 / m_i * \Sigma_{i<j}\{F_{ij}(r_{ij}, \alpha, ...)\}$ [1.3]

$$d\{r_i(t)\} / dt = v_i(t) \quad [1.4]$$

with i,j = 1 to n.

The resulting force on an atom i is expressed by an integral over all force contributions F_{ij}. Numerically this is calculated as a sum over all forces acting on each atom i (equation [1.3]). Hence, two bodies at close distance interact through this sum of force contributions in the equation of motion. To advance the atoms in space, the equation of motion has to be integrated with respect to time, once to obtain the new velocity and twice for the new position of each atom. Numerically, this operation is more efficiently carried out by approximation schemes, for instance using finite difference operators and the so-called Verlet or Stoermer algorithm [ALL 87, HOO 91]:

Verlet algorithm: $r_i(t+\Delta t) = r_i(t) + \Delta t * v_i(t) + 1/(2*m_i) * \Delta t^2 * F_i(t)$ [1.5]

$$v_i(t+\Delta t) = v_i(t) + \Delta t /(2*m_i) * \{F_i(t+\Delta t) + F_i(t)\} \quad [1.6]$$

with i = 1 to n.

With the present positions ($r_i(t)$), velocities ($v_i(t)$) and forces ($F_i(t)$), first the new positions and forces at time t+Δt and then the new velocity can be calculated. Given the equations of motion, forces and boundary conditions, i.e. knowing the current mechanical state, it is possible to simulate the future behavior of a system. Mathematically, this represents an initial value problem. A reasonable distribution of the initial velocities can be obtained from the Maxwell-Boltzmann distribution function.

The dynamic development of the atomic system as a whole determines the instantaneous kinetic state of the system. By relating the average kinetic energy of the atoms (with average velocity v), i.e. their micromechanical state, to the thermal energy of the system, which is the thermodynamic state, the gas kinetic definition of the system temperature is adopted. From equation [1.7], the temperature T of a 3D system of atoms can be directly observed or, for a given reference temperature, the kinetic energy in the system can be controlled (for details see [ALL 87, HOO 91]).

$$E_{kin} = 1/2 * m * 1/n * \Sigma v_i^2 = 3/2 * k_B * T = E_{therm.} \quad [1.7]$$

with i = 1 to n.

Since the initial choice of the atom configuration is more or less idealistic, i.e. artificial, it does not fit into the Maxwell-Boltzmann distribution from the energetic point of view. The whole system needs to pass through an initial equilibration phase, during which the atom configuration adjusts to the invariants of the system, e.g. total system energy, volume, pressure and/or temperature, and thereby also to the boundary conditions.

1.2.4. *Boundary conditions*

Boundaries are an intrinsic, vital part of models. Thermodynamic properties are thought of as characterizing "bulk" matter, which represents enough material so that surface effects and fluctuations can be ignored. To decrease the influences of boundaries, the system size needs to be chosen to be "big enough" [HOO 91]. However, the fulfillment of this weak requirement is generally limited by the available CPU power and time.

In addition to the option of free surfaces, which would result in a particle cluster in free space if applied to all axes of a Cartesian coordinate system, basically two types of boundaries are common in MD simulations: fixed and periodic boundaries. The simplest type, in terms of realization, is the fixed atom boundary which confines all freely propagating atoms inside a closed box of non-moving atoms or provides support for them at one or more sides. It is simply realized by taking away the dynamics of such boundary atoms, but keeping the interactions with the freely moving atoms. The consequences of such infinitely hard boundaries for the simulation can be significant as no energy can be passed through the boundary and phonons will be reflected at it. The sole use of hard boundaries represents a poor representation of the surrounding environment/material. Some of the negative effects of hard boundaries can be corrected by placing thermally controlled atom layers between freely moving atoms and a hard boundary [BEL 91, SHI 92].

Periodic boundary conditions (PBCs) were introduced to avoid the hard boundary reflection and allow us to study bulk and bulk/interface structures without the strong boundary influence in small models (see [ALL 87, HOO 91]). It is imagined that the bulk of the material is made of many similar systems along the axis perpendicular to the periodic boundary plane, i.e. there are no surfaces along this axis. The system reacts as if there are identical systems at both sides of the PBC, exposed to the same conditions and changes (see Figure 1.3). In practice, the system is connected to itself, and atoms at one side interact with atoms on the other side and form a continuous structure. If deformation in the system requires an atom to slip across the PBC, it transfers from one side of the model to the other. Figure 1.3 shows a sketch of an indentation model (triangular indenter on the top of a work-piece), where a one-axis PBC is considered perpendicular to the horizontal axis.

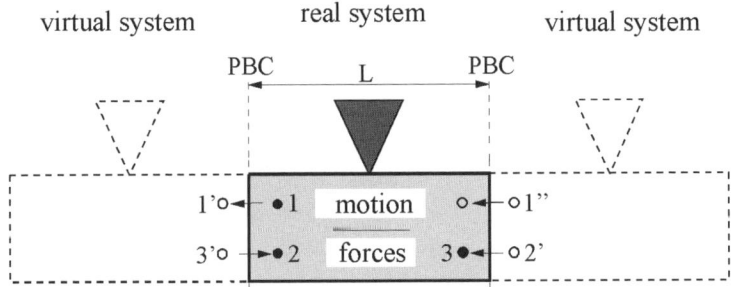

Figure 1.3. *Motion through (atom 1) and forces at (atoms 2-3) a PBC*

A consequence of periodic boundaries is that energy and phonons are not reflected, but travel through the system by means of the PBCs. One or two-axis PBCs can be employed where symmetry axes are available and the lattice structure allows an undisturbed bonding through the PBC planes. Additionally, a deformation compatibility across a PBC has to be fulfilled by an appropriate alignment of preferred slip systems relative to the PBCs, in order to avoid artificial deformation patterns.

The following results were all obtained by using 3D MD models, EAM potential functions and PBCs in one or two axes, even if the width of the underlying MD model was only a few lattice constants wide, following the approach of the orthogonal cutting process condition.

1.3. Design and requirements for state-of-the-art MD cutting process simulations

The fundamental part of a material removal process is the relative motion of two interacting bodies, where one is carrying out work, usually the tool, upon the other one, the workpiece. Therefore, the MD model needs to include, at least, the surfaces of these two interacting bodies in the area of contact and a sufficient portion of matter (see Figure 1.4). Hence, full 3-axis PBCs are not applicable since the surfaces have to be along one axis. Furthermore, a relative motion between both bodies needs to be applied to account for the cutting speed. The process requires that cutting, thrust and tool forces are balanced or accommodated at the system boundaries in order to measure, for instance, tool forces or to avoid unintended translational and rotational motion by the tool or the workpiece as a whole.

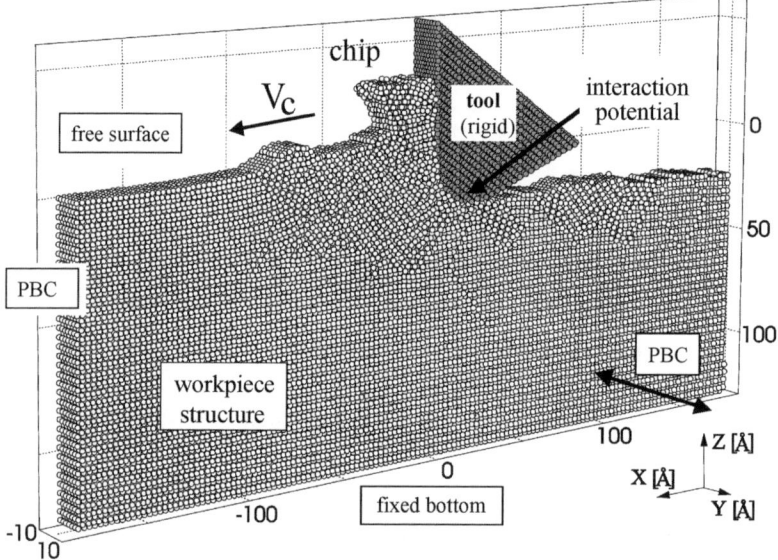

Figure 1.4. *Design and boundary conditions of a 3D MD cutting model*

In most cases the cutting process simulations focus on the effect of the material removal process on the workpiece structure, for which the tool is often modeled rigid in order to reduce the complexity of the simulation as a whole. By exerting work with the hard tool tip upon the workpiece, energy is added to the workpiece, whereupon its temperature would rise. Implementing the earlier mentioned thermally controlled atom layers around the outer boundaries of the workpiece allows us to control its temperature by drawing away energy to the non-modeled area of an imagined larger workpiece volume [BEL 91, SHI 92]. Although in the literature results from MD simulations of the machining of brittle semi-conductor and of ductile polycrystalline materials have also been presented ([GLO 93, GLO 95, REN 95-1, REN 95-3, SHI 94]), all the following figures, like Figure 1.4, show results of cutting or machining process simulations where hard tool tips move on {001} surfaces along <110> directions of fcc copper crystals.

The machining operation represents a massive deformation process. In order to reduce interactions with the boundaries and to allow for sufficient elastic deformation, a reasonably large-sized model has to be determined by tests or on the basis of experience. Experience further shows that 2D MD models have little meaning, unless the process in question is restricted to a plane and all important information can be accounted for in such a model. In most cases a 3D model or at least a semi- or quasi-2D model is the better choice, since material properties and structures are significantly better represented.

The relative motion for the material removal process can either be applied to the tool tip model or the boundary atoms of the workpiece (the freely moving atoms follow automatically). When the tool tip comes into contact with the workpiece surface, the deformation causes material to be piled up in front of the tool tip whereupon gradually a chip is formed during the course of process simulation. To reach constant cutting conditions it is necessary to observe the process and its quantities for a sufficiently long time as it will be discussed in the next chapter in more detail. For both requirements, sufficiently large model sizes as well as long process observations, the use of fast and large computing systems, like parallel-processor computers, is as desirable for MD as fast algorithms. Regarding fast algorithms, the basic idea of the introduced techniques is to reduce the calculation of forces and stresses to those that are necessary. In principle, the force calculation is at least an N^2 operation, even for simple pair potentials. Methods to speedup the calculation, like the so-called book-keeping technique, the next-neighbor-cell method [ALL 87] or the use of tabulated force functions [REN 95-1], employ static or dynamic tables at different levels of the calculation procedure. Using these techniques, the program codes become significantly more complex, but change the dependence of the calculation time on the system size from an N^2 relation to an N^1 relation. Further details about these techniques can be found in the literature [ALL 87, REN 95-1].

1.4. Capabilities of MD for nanoscale material removal process analysis

For the analysis of the MD process simulation results, micromechanical state variables of all atoms and the thermodynamic state variables of the system as a whole are accessible throughout the simulation. While some of the state variables are directly available, like coordinates and velocities ("micromechanical state"), others are only available through statistical mechanics, which need to be calculated as system quantities like the thermodynamic state variables heat, temperature and pressure.

Some of the capabilities of MD for analyzing nanoscale cutting processes by simulation will be demonstrated on the basis of application examples. For this purpose results of orthogonal cutting process simulations of a ductile, single-crystalline copper will mostly be presented, although MD cutting and machining simulation results of brittle as well as ductile polycrystalline materials can be found in the literature [GLO 93, GLO 95, REN 95-1, REN 95-3, SHI 94]. If not otherwise stated, all the following figures show results of cutting or machining process simulations where hard tool tips move on {001} surfaces along <110> directions of fcc crystals.

1.4.1. *Analysis of microstructure and deformation*

The direct availability of the atom coordinates in MD calculations allows us to visualize the instantaneous positions and to track the motion of the atoms individually as well as a whole. This information is not only useful for so-called snapshots of atom arrangements as shown in Figure 1.4, through which the progress of process simulation can be monitored, but it can be used for dynamic analyses on the basis of animations as well. For this purpose the radii of the atoms are often chosen more in order to enhance a certain aspect or view of a specific arrangement than representing realistic atom sizes. Using instantaneous coordinates rather than averaged positions, due to the vibration of the atoms, does not cause a significant error, because usually the maximum vibration amplitude of an atom in a stable MD calculation is less than 1% of the minimum bonding length.

The model in Figure 1.4 was employed to study the chip formation process and the surface generation at a cutting edge in nanoscale cutting. By employing 1D PBC along the y axis, the orthogonal cutting condition reduced the model to a quasi-2D type with a small width, which makes it possible to correctly model the 3D fcc crystal structure of the copper workpiece as well. The model contained 71,000 work atoms and 11,000 tool atoms. Atoms at the bottom and to the very left and right hand side of the model represent fixed boundaries, but were shifted with cutting speed to create the relative motion between workpiece and tool. Atoms in the layers next to these hard boundary atoms had thermostat properties and controlled the temperature at the workpiece boundary.

Since for diamond cutting of copper no plastic and no significant elastic deformation of the tool was expected, calculating its internal structure was not of special importance, but its shape and surface structure were relevant for the tool-work interaction. Its surfaces were formed by preferred diamond/fcc cleavage planes and used as rake and clearance faces. The edge radius was not chosen atomically sharp (here 2 nm), in order to consider limited minimum edge radii, because of surface stresses [IKA 77]. To further provide a reasonable tool-work contact the tool structure had the atomic density of diamond. Although the tool was modeled as a hard body with collectively moving atoms, i.e. no interaction within the tool, the interaction potential between tool and work atoms needed to be specified. Data for the diamond/copper interaction based on a pair potential function were found in Shimada *et al.* [SHI 93]. The cutting forces were calculated as reaction forces at the tool due to its infeed motion into the workpiece surface. The work atom interactions are described by Ackland's EAM potential for copper [ACK 87]. The cutting speed was restricted to 100 m/s and 50 m/s. Lower speeds were not practical to simulate due to computational limitations.

Figure 1.4 shows a single-crystalline structure, which moved towards the cutting tool, whereupon material is deformed in front of the tool tip, the chip generation is initiated and dislocation loops can be identified at the generated workpiece surface. Figure 1.5 shows a so-called deformation graph in a 2D projection of the same model and progress of simulation. It shows areas of plastic deformation, dislocations and large elastic deformations in the subsurface region. The method is based on horizontal and vertical connections between initial-neighbor atoms. Deformations appear as sharp equilateral folds in neighboring layer lines within the otherwise rectangular structure or by narrowing mesh spacing as in the case of strong elastic deformation. For large displacements between initial-neighbor atoms, the bond was considered to be broken and was no longer drawn. In this way, highly deformed areas, like the chip and the newly generated surface, show few initial-neighbor lines.

Deep running dislocations, observed in 2D MD cutting simulations if pair potentials are being used ([REN 95-2, REN 96]), could not be confirmed by employing this 3D model and the better EAM potential. This model predicts intensive plastic deformation at the generated surface with a thickness of only a few atom layers. At the same strain, a 2D model always predicts larger dislocations than its 3D counterpart. The cutting process changes drastically when changing the ratio of cut depth to cutting edge radius from 0.5 to 1.0, also depending on the crystalline orientation of the work. At a ratio of $a_e/r_\beta=1.0$, the tool begins to use its rake face more for the chip formation. With the increase in cut depth, the portion of twin dislocations in chip formation increases over dislocation slipping. Such twinned areas can be seen ahead of the tool and the chip. The lower energy requirement for twinning makes the chip removal process at larger cut depths more efficient and the cutting forces only increase proportionally.

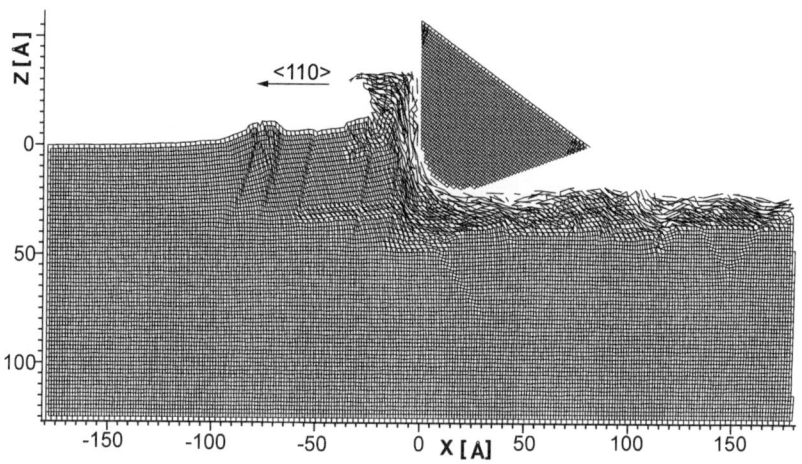

Figure 1.5. *Deformation graph, view <110>*

Figure 1.6 shows two subsequent 2D views of microstructural snapshots of the same model as in Figures 1.4 and 1.5, but at a smaller cut depth than before. Analyzing the changes in the microstructure deepens the understanding of the involved mechanisms of material removal for specific cutting conditions. In Figure 1.6 areas with different crystal orientation are separated by lines and slip lines drawn for identified dislocations. The cutting process right before state a) was characterized by a growth of the pre-deformation area in the cutting direction (ahead of the chip) without an increase in chip length. Until state b) this process had stopped, the pre-deformation area decreased while the chip grew in size. The microstructural plots show a change of the crystal orientation in the chip root area that supports either the deformation away from the chip root (a) or the chip formation in the chip root area.

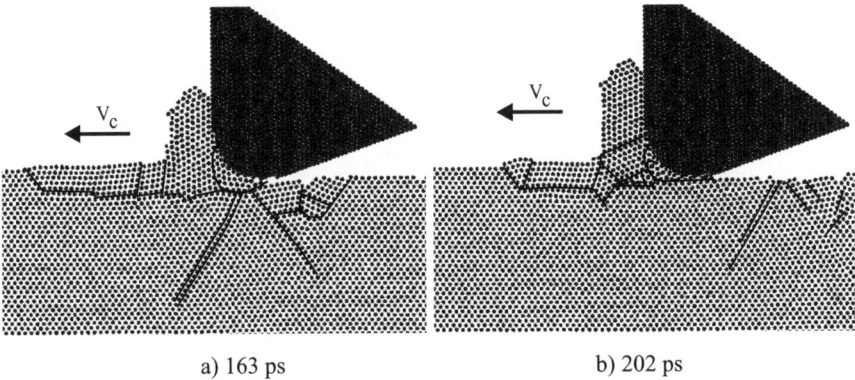

Figure 1.6. *Microstructural changes in the chip root area during chip formation*

1.4.2. *Obtaining cutting forces, stress and temperature*

When modeling systems of discrete particles and observing their progression over time, statistical mechanics provides a basis for the analysis and description of the behavior of such systems. It has been demonstrated for a Gibbs microcanonical ensemble that taking time averages is in statistical agreement with taking phase-space averages and that the numerical quality of results in MD can always be improved by longer calculations [HOO 91].

As mentioned before, cutting forces can be calculated as reaction forces at the tool due to its relative motion during contact with the workpiece atoms. For every time step Δt (see equation [1.3]), which is usually in the range of a few femtoseconds (10^{-15} s), the force contributions of the workpiece atoms interacting with the tool are integrated (see equation [1.5]). The dynamic character of such a system with a large number of degrees of freedom, i.e. all the freely moving workpiece atoms, emerge as fluctuations of derived, non-constant quantities. Figure 1.7 shows the course of the cutting force radio for the simulation in Figure 1.6.

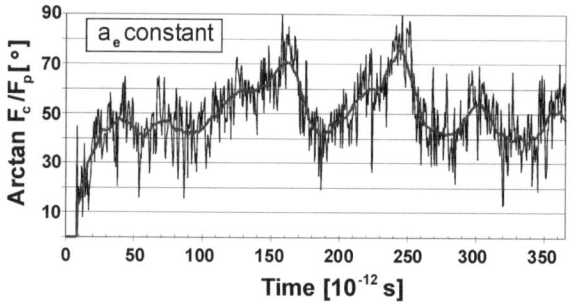

Figure 1.7. *Course of the cutting force ratio Fc/Fp*

The instantaneous tool forces, which will be newly calculated for every time step, fluctuate intensively. Calculating a moving average of the force ratio over 1,000 time steps for example cancels out the fluctuations and leads to a smooth course. After overcoming the equilibration phase (no cutting forces), the first tool/workpiece contact was made and the force ratio changed to an average value of 1 (=arctan 45°) over a period of about 20 pico seconds (10^{-12} s). Besides smaller maxima and minima during the observed total process simulation time, two gradually developing maxima in the course of the force ratio appear at about 165 pico seconds and 245 pico seconds. Figure 1.6 sheds light on the microstructural process which is related to the observed course of the force ratio.

Detailed information about the distribution of stresses and temperature in nano- and microscale cutting are of great interest for science and the manufacturing industry. So far most of the MD results of cutting process simulations were presented as atomic, discontinuous sets of instantaneous data at individual atom sites, such as snapshot atom positions, relative displacement and instantaneous atomic temperature. Besides the limited meaning of instantaneous atomic temperatures and stresses, looking at such large sets of 10,000, 100,000 or even millions of pieces of data is not practical from a point of view of efficient data analysis. Furthermore, it makes any attempt to compare MD results with, for instance, results from continuous mechanics difficult if not impossible. Taking advantage of the possibility of improving the quality of local values by calculating them as time averages over a sufficiently long period of time provides the means to obtain a deeper insight of the model and the simulated process. Thus, aiming at macroscopic thermodynamic properties, suitable time intervals for averaging these properties have to be identified. Simulations showed that an average over about 1,000 time steps led in some cases to sufficiently stable mean properties, but still provides a certain time resolution in order to study details of the process. Considering the basics of MD and the physical nature of these quantities, the results can now be represented in a form of gradual distributions as so-called contour plots, with a certain resolution in space as well as in time. The representation of stresses and temperature in terms of continuous distributions allows a direct comparison of continuous mechanic results and MD results.

In Figure 1.8 the calculated distribution of the maximum shear stress of the orthogonal cutting process already shown in Figure 1.4 and Figure 1.5 is given. With the help of this distribution it is possible to determine where stress concentrations occur, how much the crystal structure influences the stress distribution, as well as the material removal process, and what the differences are in comparison to macroscopic, continuous mechanical processes.

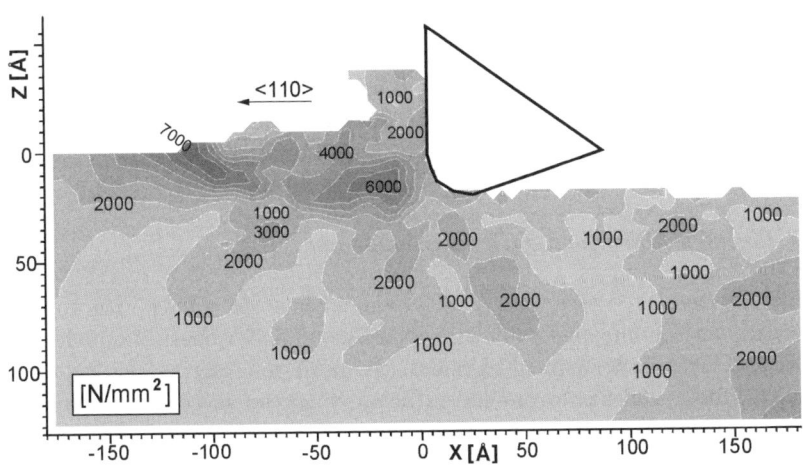

Figure 1.8. *Shear stress distribution in an orthogonal 3D MD machining model*

They also allow us to determine how deeply the process influences the workpiece and where new dislocations can occur, since areas of high shear stress are potential sources for formation or extension of dislocations. Similarly, temperature distributions can also be calculated by adapting equation [1.7] to local volumes and calculating time averages. As it will be shown in the following section (Figure 1.12b), MD cutting process simulations without the consideration of fluids lead to approximately concentric temperature distributions, in which the hottest areas are the chip area and the chip root area. Right at the tip of the tool, the material is deformed at a high stress level, whereby a lot heat is generated. Hence, the high temperature areas extend from the chip under the tip of the tool, as one important source of heat generation, to the areas of shearing. It should be noted here that modeling the tool by rigid, thermally inactive atoms does not enable the tool to conduct any heat. Therefore, the tool acts like a thermal isolator, which further supports a concentration of heat in the chip.

Regarding the temperature distributions in MD cutting simulations, it should be further noted that in most of the published works, only the thermal conductivity through phonons is considered. Conductivity by electrons is neglected in such cases, even though it is one order of magnitude larger than that of the phonons. Hence, the presented temperature levels as well as the local gradients would actually be lower than shown. New algorithms were developed to describe thermal conductivity more accurately by considering both the electron and phonon conductivity [CAR 94].

1.5. Advances and recent developments in material removal process simulation

In this chapter two aspects of advances and recent developments in material removal process simulation using MD will be explained in more detail. They are the possibility of carrying out complete 3D surface machining simulations and the consideration of fluids.

1.5.1. *Complete 3D surface machining simulation*

For abrasive processes, the model requirements are higher than for cutting processes, since orthogonal symmetry is not given and a quasi-2D model, like in Figure 1.4, cannot be applied. Besides the need to describe the geometry of abrasives, the model has to provide sufficient space for the deformation and chip formation of the 3D material removal process. Figure 1.9a shows a snapshot of a molecular dynamics simulation to study material pile-up and chip formation in abrasive machining as a function of shape and orientation of the abrasives.

The simulation in Figure 1.9 considered two hard pyramidal grits with diamond structure and two different orientations. The figure shows an advanced state of simulated 3D grinding using a model with more than 100,000 copper atoms (the workpiece height was 6 nm). In several ways the simulation represents a high-end state-of-the-art MD simulation of the grit/workpiece contact as the interactions were based on an EAM potential function [ACK 87] and the model considers two abrasives that cut at 100 m/s through a workpiece over its whole length. Thus, the periodic boundaries (for both directions of the horizontal plane) lead to complete groove formation by the grits in cutting direction and describe a model setup with multiple grit/workpiece contacts like in grinding (see Figure 1.9b). By repeating the complete groove generation with relative-to-the-cutting-direction shifted abrasives, the machining of the whole surface can be realized. This provides the basis for 3D surface roughness and residual stress analyses of completely machined surfaces at realistic machining speeds (common grinding speeds range from about 5 to 80 m/s and high speed grinding up to about 250 m/s).

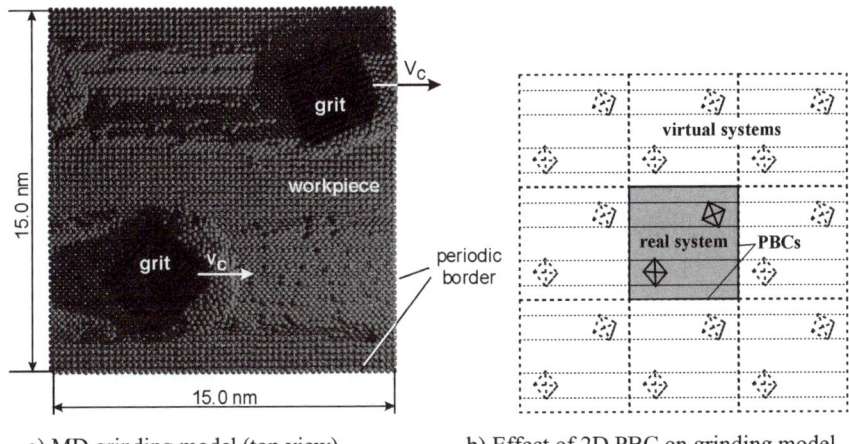

a) MD grinding model (top view) b) Effect of 2D PBC on grinding model

Figure 1.9. *Snapshot of a full length model scratching simulation with 2 hard abrasives (after 360,000 time steps, 144 ps) [BRI 06]*

Thorough analyses of the chip formation, the elastic and plastic response of the workpiece and process quantities in MD simulations have revealed clear and consistent effects. As [REN 96] and [SZE 93] show, for example, the machining speed has a direct influence on the microscopic material removal process and the chip formation in MD simulations. The results suggest that the sensitivity of the simulation results on the machining speed is less strong than observed in experimental investigations. A possible reason for this effect is that the implemented boundary conditions and model settings have a strong impact on the dynamics of the finite process model. However, significant changes in magnitude of the machining speed lead to significant changes in chip shape and formation mechanism. Further quantities of the process are effected, due to an increasing localization of the deformation process at high speeds. More direct effects on the process simulation results show the depth of cut, the grit orientation and the cutting edge radius. Hence, it is possible with MD to simulate the influence of grain shape and orientation on the efficiency of abrasive processes. On the basis of bigger MD models, it will be possible to determine the energy dissipation by a direct analysis of elastic and plastic work and the microscopic mechanisms that determine the surface roughness in nanoscale machining processes.

1.5.2. *Consideration of fluids in MD cutting simulation*

Most state-of-the-art material removal process simulations focus on the material removal mechanisms, chip and surface generation [MAE 95, KOM 99, REN 95-2, SHI 93, SHI 94]. Besides partially strong idealizations of the tool and workpiece properties as well as the direct contact and boundary conditions, so far fluids have not been included in the environmental descriptions of workpiece and tool in MD-modeling. Hence, such an environment represents high vacuum conditions with no heat convection to an atmosphere or coolant. Therefore, an extension of MD machining process simulation has been proposed by considering fluids together with the tool tip and the workpiece [REN 05]. The extension of MD machining process models by molecular gas and fluid dynamics (see [HOO 91, RAP 95]) provide an opportunity for enabling a complete energy balance and for investigating the impact of adsorption and reaction layers at the workpiece surface and their contribution to the contact tribology beyond dry-machining at high vacuum. For this purpose the spaces above workpiece and tool surfaces need to be filled with particles that follow fluid particle dynamics.

Figure 1.10 shows some details of the design of the 3D simulation cell, which is identical to those in Figures 1.4 to 1.8, except for the atomic fluid and the cut depth, which was chosen as $a_e = 0.5$ nm (like in Figure 1.6). The cutting speed was 100 m/s, also as in previous figures. The fluid filled the space above the workpiece surface and around the tool. As a modeling approach, a simple, atomic, non-reacting fluid was considered. The data for the fluid were derived from those of water with a density of 1.0 g/cm^3 and molar mass of 18 g, which results in a fluid mass of $2.988*10^{-26}$ kg and average bonding distance of 0.3154 nm. The atomic approach ignores the rotational molecular moments of inertia and reduces its mass to the center of gravity. The data for the molecule-molecule interaction energy of water were found in [RAP 95]. The fluid-fluid interactions as well as the tool (hard)-workpiece interactions were calculated on the basis of the Lennard-Jones (LJ) potential function. The wetting of the tool (fluid-tool) and workpiece surface (fluid-workpiece) were also described by the fluid-fluid potential function that represents a weak interaction like hydrogen bonding in water, but no chemical reaction. As the consideration of a fluid in the cutting process simulation was the focus of this work, internal tool dynamics were not considered here. For the inner workpiece reactions an EAM was applied (copper [ACK 87]). The data for the tool-workpiece interaction represent a friction contact in diamond cutting with suitable lattice constant and cut-off distance [IKA 91].

Nanoscale Cutting   21

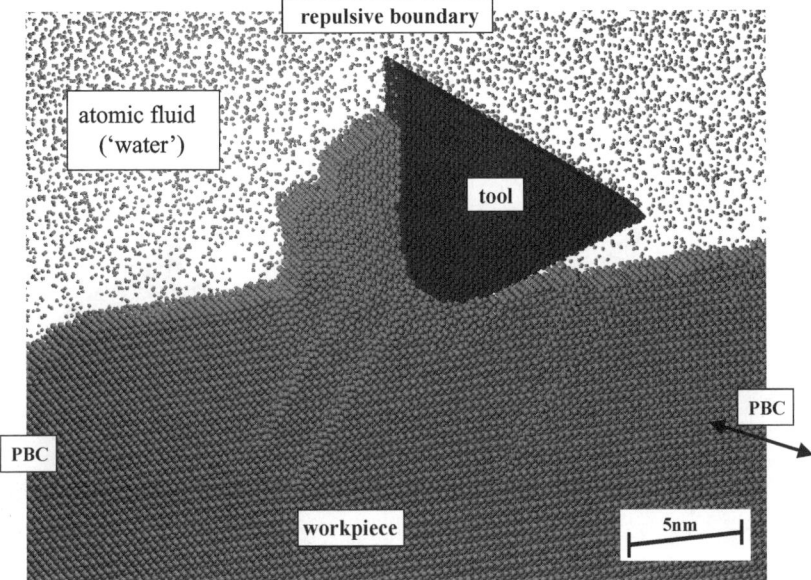

Figure 1.10. *Snapshot of an MD-cutting simulation considering an atomic fluid (after 600,000 time steps, 240 ps)*

By adding only the fluid to the MD model, a direct comparison with earlier results is possible and the impact of the fluid on the material removal process simulation can clearly be pointed out. Figure 1.11 shows the impact of the fluid on the chip formation and Figure 1.12 shows the effect on the temperature distribution.

In both simulations in Figure 1.11 the tool was initially not in contact with the workpiece, and followed the same trajectory and cut for about the same distance as the workpiece. With the presence of the fluid the material removal process starts earlier and the chip formation appears more efficient, since more material is piled up at the rake face of the tool. Thus, at vacuum condition (no fluid), more material is drawn underneath the tool tip, for which the workpiece must store higher elastic stress and more elastic deformation below the tool than with the fluid atmosphere. This phenomenon is closely related to the workpiece surface stresses which will be reduced by the interaction with the fluid atoms.

It is interesting to notice that the fluid density is lower at the workpiece surface, where the fluid atoms pick up heat from the heavier and thermally active (freely vibrating) workpiece atoms. At the same time, the tool atoms were not vibrating, i.e. they could not exchange heat with the fluid atoms. As a consequence, many fluid atoms adhered to the tool surface and formed a fluid layer (see Figure 1.11a).

In order to visualize the thermal effect of fluids in MD cutting simulation, the atomic fluid was added to the model in Figure 1.11b after cutting was carried out already for 202 ps without a fluid. Then the simulation was continued for some time with the fluid.

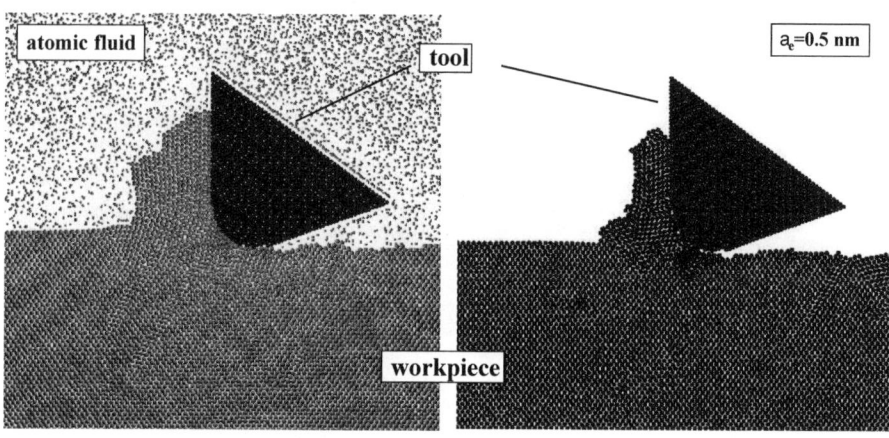

a) with fluid (after 240 ps)   b) without fluid (after 230.7 ps)

Figure 1.11. *Effect of the presence of a fluid in MD cutting simulations*

In Figure 1.12 the temperature distributions for both cases are presented, with the fluid added (a) and without the fluid (b). In Figure 1.12b a typical temperature distribution of an MD cutting simulation without a fluid is shown. The highest temperatures concentrate around the tool contact area and the chip root, but the temperatures in the chip are about the same level. A very different result was found when the fluid was added. Figure 1.12a no longer presents a high-temperature region. On average the temperature workpiece is only slightly elevated above the environmental temperature, respectively above the fluid temperature of 300 K. It should be noted that the workpiece temperature was controlled at its outer boundaries only.

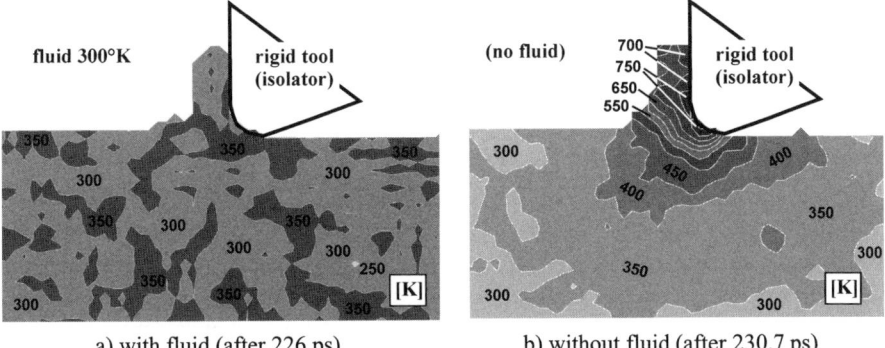

Figure 1.12. *Temperature distributions in MD cutting simulations with and without a fluid*

1.6. Summary and outlook

The aim of this chapter is to provide an insight into the basics of molecular dynamics needed for nanoscale cutting simulations, to describe its state of the art and to demonstrate some of its capabilities for cutting process analysis on the basis of a number of example applications, as well as to point out recent developments in this field.

With its well based physical concept, MD modeling allows for a detailed and comprehensive representation of material properties, where elastic, plastic materials as well as hard and brittle materials are treated within one framework. Once the atomic interaction is described, atomic contact, friction, fracture and plastic deformation results from the interaction with neighbor atoms. MD provides the means for a better insight into the essential processes of cutting at nanometer level, such as the localized material behavior in particular.

The future trends in MD process simulation clearly point towards larger 3D atomistic modeling, better material representation and longer simulated process time. Many-body potential functions for better metal representation are available for some materials. However, the further development of MD applications faces two major principle problems concerning the absolute model size, respectively the large number of atoms, and the need for describing the interactions between all elements involved. With the spreading of high-performance parallel computing facilities, the main bottleneck is shifting to efficient data processing and representation. Regarding the interactions, there is a lack of realistic potential functions for most of the typical engineering materials, in particular for compounds and alloys. This is an active field of development, combining new computational techniques with novel experiments. The progress in the development of new potential functions is supported by activities in physics, physical chemistry and material science. In a direct approach the lack of interatomic potential functions is solved by combining quantum mechanics calculation and MD. However, usually the size of such models is less than 1,000 atoms.

The state of the art in MD grinding, scratching, cutting or indentation simulation does not consider fluids. Thus, its environment represents high vacuum with no heat convection. The extension of the MD cutting process models by molecular fluid dynamics provide an opportunity for realizing a complete energy balance and for investigating the impact of adsorption and reaction layers and their contribution to the contact tribology beyond dry-machining at high vacuum. Fundamental tribology and friction investigations are already on the way and international conferences have established whole MD sessions. In recent years, many molecular modeling research groups have been formed in various fields and institutions all around the world.

1.7. References

[ACK 87] Ackland G.J., Tichy G., Vitek V., Finnis M.W., "Simple N-body Potentials for the Noble Metals and Nickel", *Phil. Magazine*, A., Vol. 56, No. 6, 1987, p. 735-756.

[ALL 87] Allen M.P., Tildesley D.J., *Computer Simulation of Liquids*, Clarendon Press, Oxford, 1987.

[BEL 91] Belak J., Stowers I.F., "The Indentation and Scraping of a Metal Surface: A Molecular Dynamics Study", in *Fundamentals of Friction: Macroscopic and Microscopic*, Eds.: Singer, Pollock, ASI Series E, Vol. 220, 1991, p. 511-520.

[BRI 06] Brinksmeier E., Aurich J.C., Govekar E., Heinzel C. *et al.*, "Advances in Modeling and Simulation of Grinding Processes", Keynote paper STC G, Annals of the International Academy for Production Engineering (CIRP), Vol. 55/2, 2006, p. 667-696.

[CAR 94] Caro A., "Electron-Phonon Coupling in Molecular Dynamics Codes", *Radiation Effects and Defects in Solids*, Vol. 130-131, 1994, p. 187-192.

[DAW 84] Daw M.S., Baskes M.I., "Embedded-atom Method: Derivation and Application to Impurities, Surfaces, and other Defects in Metals", *Phys. Rev.* B 29 (12), 1984, p. 6443-6453.

[FIN 84] Finnis M.W., Sinclair J.E., "A Simple Empirical N-body Potential for Transition Metals", *Philosophical Magazine*, A., Vol. 50, No. 1, 1984, p. 45-55.

[GLO 93] Glosli J.N., Boercker D.B., Tesar A., Belak J., "Nanoscale Plasticity in Silica Glass", *Proc. of the ASPE*, Annual Meeting, 1993, p. 111-114.

[GLO 95] Glosli J.N., Belak J., "Molecular Dynamics Modeling of Ultra Thin Amorphous Carbon Films", 4th Int. Symposium on Diamond Materials, *Electrochemical Society Proceedings*, Vol. 95(4), 1995, p. 25-37.

[HOO 90] Hoover W.G., De Groot A.J., "Large-scale Elastic-Plastic Indentation Simulations via Nonequilibrium Molecular Dynamics", *Phys. Rev.* A 10 (42), 1990, p. 5844-5853.

[HOO 91] Hoover W.G., *Computational Statistical Mechanics*, Studies in Modern Thermodynamics 11, Elsevier Science, Amsterdam–Oxford–New York–Tokyo, 1991, 313.

[IKA 77] Ikawa N., Shimada Sh., "Cutting Tool for Ultraprecision Machining", *Proc. of the 3rd. Int. Conf. on Production Engineering*, Kyoto, July 14-16th, 1977, p. 357- 364.

[IKA 91] Ikawa N., Shimada Sh., Tanaka H., Ohmori G., "An Atomistic Analysis of Nanometric Chip Removal as Affected by Tool-work Interaction in Diamond Turning", CIRP, 40/1, 1991, p. 551-554.

[KOM 99] Komanduri R, Chandrasekaran N., Raff L.M., "Orientation Effects in Nanometric Cutting of Single Crystal Materials: An MD Simulation Approach", *Annals of the CIRP*, 48/1, 1999, p. 67-72.

[LAN 89] Landman U., Luedtke W.D., Nitzan A., "Dynamics of Tip-substrate Interactions Atomic Force Microscopy", *Surface Science*, Vol. 210, 1989, p. 177-184.

[MAE 95] Maekawa K., Itoh A., "Friction and Tool Wear in Nano-scale Machining – A Molecular Dynamics Approach", *Wear 188*, 1995, p. 115-122.

[RAP 95] Rappaport D.C., *The Art of Molecular Dynamics Simulation*, Cambridge Univ. Press, Cambridge, UK, 1995.

[REN 95-1] Rentsch R , Inasaki I., "Indentation Simulation on Brittle Materials by Molecular Dynamics", in *Modeling, Simulation, and Control Technologies for Manufacturing*, SPIE Proc. Vol. 2596, Ed. R. Lumia, Philadelphia, PA, USA, 1995, p. 214-224.

[REN 95-2] Rentsch R., Inasaki I., "Investigation of Surface Integrity by Molecular Dynamics Simulation", *Annals of the CIRP*, Hallwag Publ., Berne, CH, Vol. 44/1, 1995, p. 295-298.

[REN 95-3] Rentsch R., Inasaki I., "Simulation of Single Point Machining of Poly-crystalline Metals by Molecular Dynamics (MD)", *International Progress in Precision Engineering*, Editors M. Bonis *et al.*, Compiègne, F, Elsevier, 1995, p. 347-350.

[REN 96] Rentsch R., Inasaki I., Brinksmeier E., *et al.*, "Influence of Material Characteristics on the Micromachining Process", *Materials Issues in Machining-III* and *The Physics of Machining Processes-III*, Editors D.A. Stephenson, R. Stevenson, TMS Publication, Cincinnati, Ohio, USA, 1996, p. 65-86.

[REN 01] Rentsch R., "On the On-set of Chip Formation and the Process Stability in Cutting at Atomic Level", *Initiatives of Precision Engineering at the Beginning of a Millennium*, Ed. I. Inasaki, Kluwer Acad. Publ., 10. Int. Conf. on Precision Engineering (ICPE), Yokohama, Japan, 2001, p. 274-278.

[REN 05] Rentsch R., Brinksmeier E., "Tribology Aspects in State-of-the-art MD Cutting Simulations", *8th. CIRP Int. Workshop on Modeling of Machining Operations*, Chemnitz, D, 2005, p. 401-408.

[REN 06] Rentsch R., Brinksmeier E., "Numerical Simulation of Residual Stresses at the Grain and Sub-grain Length Scale using Atomistic Modeling", Trans Tech Publ., CH, *Materials Science Forum*, Vols. 524-525, 2006, p. 517-522.

[SHI 92] Shimada Sh., Ikawa N., Ohmori G., "Molecular Dynamics Analysis as Compared with Experimental Results of Micromachining", *Annals of the CIRP*, Vol. 41/1, 1992, p. 117- 120.

[SHI 93] Shimada Sh., Ikawa N., Tanaka H., Ohmori G., Uchikoshi J., "Feasibility Study on Ultimate Accuracy in Microcutting using Molecular Dynamics Simulation", *Annals of the CIRP*, Vol. 42/1, 1993, p. 91-94.

[SHI 94] Shimada Sh., Ikawa N., Tanaka H., "Structure of Micromachined Surface Simulated by Molecular Dynamics Analysis", *Annals of the CIRP*, 43/1, 1994, p. 51-54.

[SZE 93] Shimizu J., Zhou L., Eda H., "Molecular Dynamics Simulation of Material Removal Mechanism beyond Propagation Speed of Plastic Wave", LEM21, *JSME*, 2003, p. 309-314.

[TER 90] Tersoff J., "Modeling Solid-State Chemistry: Inter Atomic Potentials for Multi-component Systems", *Phys. Rev. B*, Vol. 39(8), 1989, p. 5566-5568, and 1990, Vol. 41(5), p. 3248.

[YIP 89] Yip S., Wolf D., 1989, "Atomistic Concepts for Simulation of Grain Boundary Fracture", *Materials Science Forum*, Vol. 46, p. 77-168.

Chapter 2

Ductile Mode Cutting of Brittle Materials: Mechanism, Chip Formation and Machined Surfaces

2.1. Introduction

Traditionally, brittle materials, such as glass, ceramics and semiconductor wafer materials, cannot be machined by cutting, because cutting brittle materials under conventional cutting conditions chip formation would induce in a fracture mode, whereby fractured machined surface would be generated. However, during the 1980s, it was found that under special cutting conditions brittle materials can be machined by cutting in a ductile mode. Since then ductile mode cutting of brittle materials has been investigated by many research groups over the world.

Toh and McPherson [TOH 86] found that plastically deformed chips were formed in machining of ceramic materials if the scale of the machining operation is small (less than 1 µm depth of cut), that is, ductile mode cutting of brittle materials could be achieved if the cut depth is extremely small. Blake and Scattergood [BLA 90], who studied the precision machining of germanium and silicon using single-point diamond turning, pointed out that the critical underformed chip thickness is a pivotal parameter, which governs the transition from plastic flow to fracture along the tool nose. Puttick *et al.* [PUT 95] conducted single-point diamond turning of glass using cut depths in the order of 100 nm and achieved a surface quality corresponding to that

---

Chapter written by Xiaoping LI.

achieved by optical polishing, $R_a \approx 0.6$ nm, but the subsurface damage could also be observed under the condition of ductile regime machining. Inamura et al. [INA 97] [INA 99] conducted a molecular dynamics (MD) simulation study on crack initiation in machining of monocrystalline silicon. Their results showed that a microcrack-like defect could be initiated while cutting a silicon monocrystal at a cut depth of 1 micron. They assumed that the defect was created though the interaction between a local microscopic static stress and a global dynamic stress associated with acoustic waves. Zhang and Tanaka [ZHA 98] used MD simulation to study the atomic scale deformation in silicon induced by two-body and three-body contact sliding and found that amorphous phase transformation is the main deformation mechanism. Zhang and Tanaka [TAN 99] also simulated silicon indentation and the results showed that the inelastic deformation of silicon is solely caused by amorphous phase transformation. Cheong and Zhang [CHE 00] further indicated that there is a phase transformation from diamond cubic structure to β silicon in a small zone near the indenter tip using the MD simulation of nanoindentation of silicon. Fang and Venkatesh [FAN 98] reported that for turned silicon surfaces with the roughness value of $R_a = 23.8$ nm, mirror surfaces of 1 nm roughness were achieved repeatedly by micro-cutting, where the cut depth was 1 μm. Leung et al. [LEU 98] carried out direct machining of silicon on a precision lathe and found that in order to produce a high quality surface with roughness down to 2.86 nm, it is necessary that the machining process is in a ductile regime and the undeformed chip thickness must be less than a critical value, which depends on the machining conditions. Yan et al. [YAN 01] studied the role of hydrostatic pressure in ductile machining of silicon using a single crystal diamond tool with a large negative rake and undeformed chip thickness in the nanoscale range.

The cutting was arranged inside a high external hydrostatic pressure apparatus and the results indicated that large hydrostatic pressure is helpful to realize ductile mode cutting. Li et al. [LI 03] developed a nanoindentation method for precision measurement of the nanoscale diamond tool cutting edges. With the precise values of the diamond tool cutting edge radius measured using the nanoindentation method, Liu et al. [LIU 07] conducted cutting of silicon wafer with the undeformed chip thickness varying in a range from smaller to larger than the cutting edge radius, the results showed that the chip formation was in ductile mode as far as the undeformed chip thickness was smaller than the cutting edge radius; Arefin et al. [ARE 05] [ARE 07] investigated the cutting edge radius in relation to ductile mode chip formation and found that there is an upper bound for the cutting edge radius, beyond which the chip formation mode changes from ductile to brittle, and the machined workpiece surface and subsurface become fractured. Based on these findings, Cai et al. [CAI 08] studied the mechanism of ductile chip formation in nanoscale ductile mode cutting of silicon and reported an extremely high compressive stress condition in the chip formation zone as the mechanism of ductile chip formation in nanoscale ductile mode cutting of brittle materials. In the case of ductile mode cutting of silicon, the extremely high hydrostatic stress in the chip formation zone transforms the material in the chip formation zone from crystalline to amorphous, in which most of the bond lengths of the atoms become longer, making the material ductile.

Some researchers also studied the effect of ultrasonic vibration on machining of brittle materials. In glass cutting, it was found that the critical depth of ductile mode cutting can be increased by applying ultrasonic vibration to the cutting tool (Moriwaki et al. [MOR 92)]. Ultrasonic vibration assisted cutting as the precision machining process was carried out on optical plastics and lenses by Kim and Choi [KIM 97] [KIM 98]. The machined workpiece surface in ultrasonic vibration assisted cutting appeared to be from ductile cutting at very small cut depth. Moreover, the surface obtained using a single-crystal diamond tool was better than that obtained using a polycrystal diamond tool from the viewpoint of ductile cutting. Experimental results of ultrasonic vibration assisted cutting for optical plastics and lenses confirmed that the chips generated by ductile mode cutting were obtained at an extremely low cutting velocity of the ultrasonic vibration assisted cutting system. Komanduri et al. [KOM 01] built a 3D model to simulate and investigate the effect of different tool edge rake angles (from -60° to 60°) on nanoscale cutting of monocrystalline silicon with perfectly sharp tool edge and the simulation of silicon cutting showed an extrusion-like chip formation process without a specific crystallographic preference. The authors explained that this appeared to be due to a phase transformation from diamond cubic structure to β silicon under a hydrostatic pressure and consequent densification of the material. In addition, Glardon and Finne [GLA 81], Yan et al. [YAN 03], Li et al. [LI 05] and Cai et al. [CAI 07a] investigated the wear of single point diamond tools in turning of brittle materials.

In this chapter, the mechanism, chip formation and machined workpiece surface of ductile mode cutting of brittle materials are described in detail.

2.2. The mechanism of ductile mode cutting of brittle materials

2.2.1. *Transition of chip formation mode from ductile to brittle*

A transition of chip formation from ductile mode to brittle mode was observed in groove cutting on tungsten carbide. The grooving tests were carried out on a CNC lathe (Mori Seiki SL-35) using a solid CBN tool. The workpiece surface was set at an angle inclined to the tool cutting velocity direction to achieve a variation in the cut depth from zero to 10 µm during each grooving test. The average surface roughness of tungsten carbide workpiece used in this study, R_a, was less than 0.2 µm. The experimental configuration of the inclined plane grooving test is shown schematically in Figure 2.1. The tungsten carbide workpiece was fixed on the outer cylindrical surface of a disk fixture fastened by an L-shape clamp, which was held by the 3-jaw hydraulic chuck of the CNC lathe. The tungsten carbide sample's surface was inclined to the assumed working plane of 40 µm/12.7 mm on the outer diameter surface of the disk fixture, which was carefully pre-ground and well set for the grooving experiment. Indexable triangular CBN inserts were used as the grooving tool. For the tool geometry, the rake angle was 0°, the clearance angle was 11°, the cutting edge inclination was 0°, the tool included angle or point angle was 60° and insert nose radius was 0.4 mm. The radius of the CBN arc cutting edge was 6.2 µm, measured from a SEM photograph of the tool edge cross-section, and the tool rake angle formed by the cutting edge chamfer was -29.6°. The diameter of the fixture together with the tungsten carbide workpiece was 203.4 mm. The cutting speed v was 144 m/min. Six grooves were obtained in the grooving test.

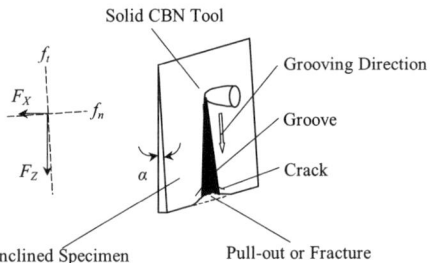

Figure 2.1. *Schematic illustration of the experimental setup for the grooving tests. (Source: [LIU 04] National University of Singapore)*

Figure 2.2 shows the machined workpiece surfaces and chips from grooving of tungsten carbide as the undeformed chip thickness increased from zero to a large value. There was a critical value for the undeformed chip thickness (at the A-A location). The machined workpiece surface was smooth and the chips generated were continuous when the undeformed chip thickness was smaller than the critical value, indicating a ductile mode chip formation. On the contrary, when the undeformed chip thickness was larger than the critical value, the machined workpiece surface was fractured and the chips generated were continuous, indicating a brittle mode chip formation. A total of 8 cuts were conducted. The average of the critical value for the undeformed chip thickness was 4.7 µm, smaller than the radius of the tool cutting edge, which was 5.2 µm.

Ductile Mode Cutting of Brittle Materials 31

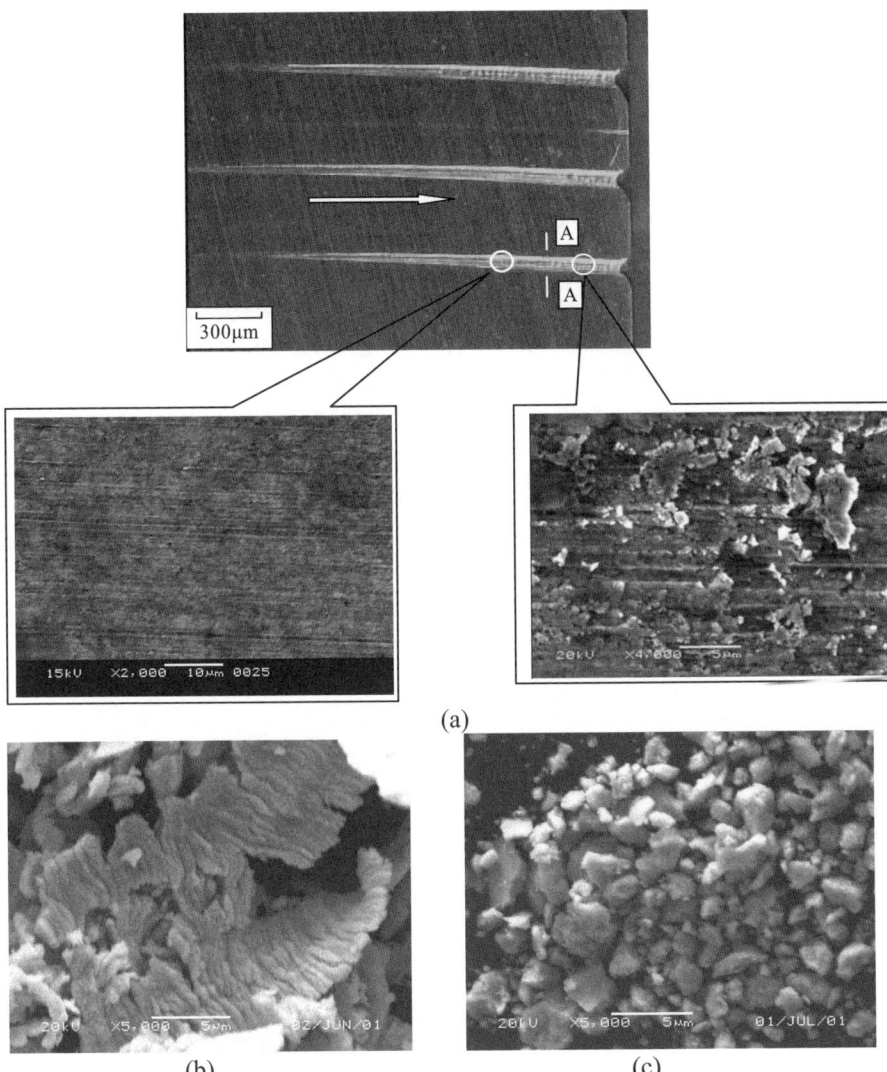

Figure 2.2. *SEM micrographs of the machined workpiece surfaces and chips from grooving of tungsten carbide with the depth increased from zero to a large value: (a) the micrographs for the machined surfaces before and after the cut depth reached a critical value; (b) the continuous chips generated from the grooving before the cut depth reached a critical value; (c) the particle chips generated from the grooving after the cut depth reached a critical value. (Source: [LIU 04] National University of Singapore)*

2.2.2. MD modeling and simulation of nanoscale ductile mode cutting of silicon

A 3D MD simulation model was employed to investigate the mechanism of ductile mode cutting of silicon. The reason for using MD simulation in this investigation was based upon the fact that nanoscale cutting involves workpiece deformation in only a few atomic layers near the workpiece surface. Figure 2.3(a) shows a schematic of the MD model and Figure 2.3(b) shows an output model of the MD simulation system. In the model, the workpiece is monocrystalline silicon and divided into three different zones: boundary atom zone, thermostat atom zone and Newtonian atom zone. The boundary atoms are fixed in positions to reduce the boundary effects and maintain proper symmetry of the lattice. The motion of the Newtonian atoms is determined by the forces produced by Newton's equation of motion. The thermostat atoms, which are used to simulate heat conduction, are arranged to surround the Newtonian atoms and make the boundary temperature close to the environmental temperature. The diamond cutting tool is assumed to be infinitely rigid such that the relative positions between tool atoms are unchanged during the cutting process. To reduce the boundary effect, the model scale should be large. As a result, the calculation time would be enormous. To avoid this problem, a periodic boundary condition is maintained along the direction perpendicular to the cross section, as shown in Figure 2.3(a).

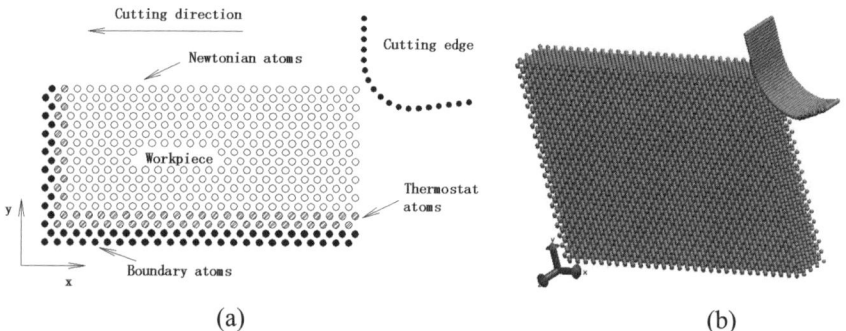

Figure 2.3. *The model for MD simulation of nanoscale ductile mode cutting: (a) a schematic of the MD model, (b) a model output from the MD simulation system. (Source: [CAI 07c] National University of Singapore)*

2.2.3. The mechanism of ductile mode chip formation in cutting of silicon

Through a MD simulation study of nanoscale ductile mode cutting of monocrystalline silicon wafer, it was found that in the chip formation zone the silicon phase transformed from monocrystalline to amorphous, forming the basis for ductile mode chip formation. In the simulation, the tool cutting edge radius R was 3.5 nm and the undeformed chip thickness was 2.8 nm ($a_c < R$). As shown in Figure 2.4, during the cutting process, the cutting geometry of the tool cutting edge radius was larger than the workpiece undeformed chip thickness, resulting in an extremely large compressive stress in the chip formation zone, which transformed the monocrystalline silicon (see insert (a) of Figure 2.4) into amorphous silicon (see insert (b) of Figure 2.4).

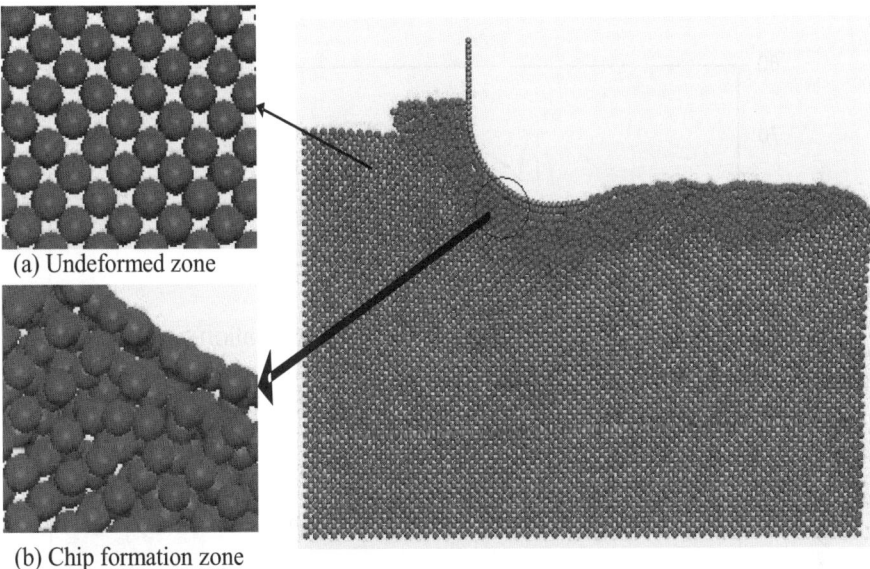

Figure 2.4. *Silicon phase transformation result from MD simulation nanoscale ductile mode cutting, showing silicon phase transformation from monocrystalline in the underformed zone (see insert (a)) to amorphous (see insert (b))*

In the amorphous phase shown in insert (b) of Figure 2.4, the atomic structure of the silicon was changed, with the interatomic bond length of 2.35 Å for silicon in the undeformed zone, which is consistent with the theoretical value of covalent bond length between the atoms in silicon, changed to interatomic bond lengths varying in a wide range in the chip formation zone, as shown in Figure 2.5. The values of the interatomic bond lengths in the chip formation zone had two peaks, at 2.455 Å and 2.61 Å, respectively, indicating silicon phase transformation from diamond cubic structure to β silicon in the chip formation zone. The β phase of silicon is a metallic body-centered tetragonal structure and can lead to plastic deformation. Moreover, in both β silicon and amorphous phase silicon in the chip formation zone, nearly all the bond lengths became longer than those of original monocrystalline silicon. According to Gao et al. [GAO 03], for pure silicon, the hardness is inversely proportional to $l^{4.5}$, where l represents the interatomic bond length. Therefore, a slight increase in the bond lengths will greatly soften the silicon atom groups. Since nearly all the bond lengths of the silicon material in the chip formation zone became longer than the normal bond length (see Figure 2.5), the material in the chip formation zone became softer than the original silicon. As a result, the silicon workpiece material in the chip formation zone was deformed in a form of plastic flow – ductile mode chip formation.

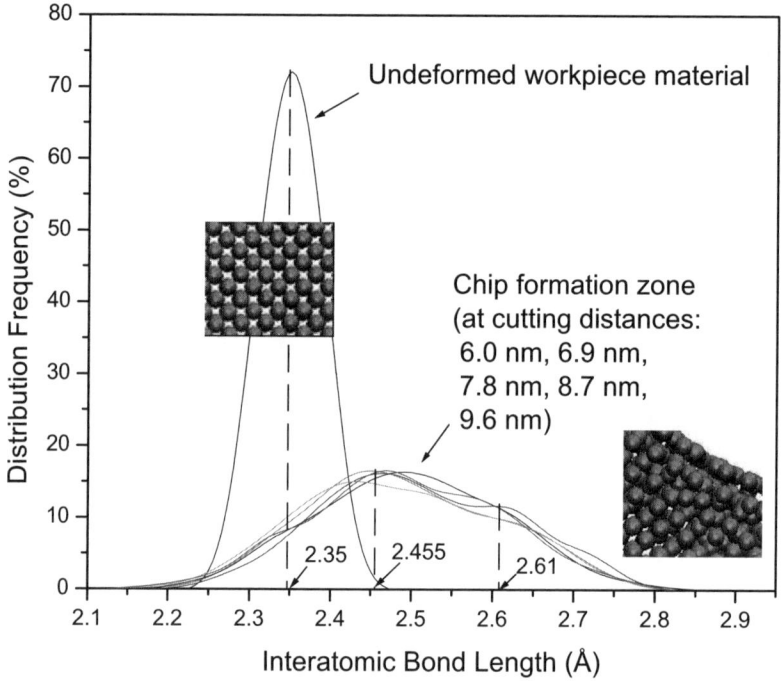

Figure 2.5. *Comparison of distribution frequency of interatomic bond length in the undeformed silicon workpiece material with that in the chip formation zone at different cutting distances*

The silicon phase transformation from crystalline to amorphous in the chip formation zone could be caused by the extremely high hydrostatic pressure in the zone, given by the tool cutting edge. According to the simulation results, the average hydrostatic pressure was around 12 GPa, which is high enough for the phase transformation [ZHA 99] [CHE 01].

2.3. The chip formation in cutting of brittle materials

2.3.1. *Material deformation and crack initiation in the chip formation zone*

The crack initiation in nanoscale cutting of monocrystalline silicon has been studied through MD modeling and simulation. In the simulation, the tool cutting edge radius R was fixed at 4.0 nm, and three cuts were performed, at the undeformed chip thickness of $a_c < R$ ($a_c = 3.2$ nm), $a_c = R$ ($a_c = 4.0$ nm) and $a_c > R$ ($a_c = 4.5$ nm) respectively.

It can be seen from Figure 2.6(a) that when the undeformed chip thickness was smaller than the cutting edge radius, the largest deformation took place at the surface of the workpiece. However, when the undeformed chip thickness was equal to or larger than the cutting edge radius, there was a peak deformation zone in the chip formation zone corresponding to the connecting point between the tool edge arc and the rake face, as shown in Figures 2.6(b) and (c), respectively.

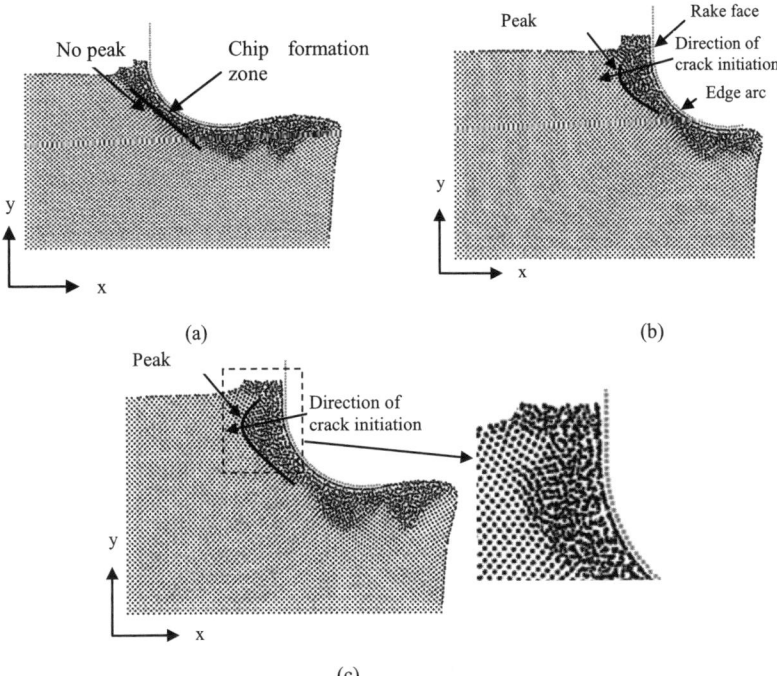

Figure 2.6. *Workpiece deformation in the chip formation zone when $R = 4.0$ nm and the undeformed chip thicknesses were: (a) $a_c = 3.2$ nm, (b) $a_c = 4.0$ nm and (c) $a_c = 4.5$ nm. (Source: [CAI 07b] National University of Singapore)*

2.3.2. Stress conditions in the chip formation zone in relation to ductile-brittle mode of chip formation

Consider the stress conditions in the workpiece in front and next to the peak deformation zone in the chip formation zone, as shown in Figure 2.7, where the σxx and σyy represent the normal stresses in the x and y directions. Figure 2.8 shows the variations of the stresses, where the simulation steps indicate the tool advancing against the workpiece. It can be seen that as the undeformed chip thickness increased from smaller to larger than the tool edge radius, σxx was positive (compressive stress) for all cases of undeformed chip thickness. The stress in the direction nearly perpendicular to the direction of crack initiation, σyy, however, decreased as the undeformed chip thickness increased. In particular, the stress state changed from compressive to tensile (negative).

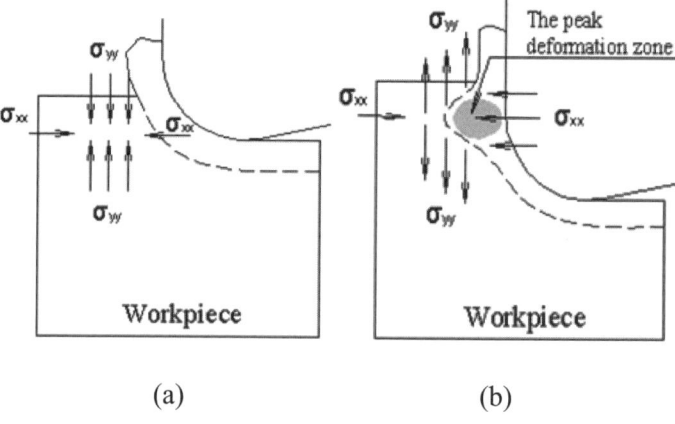

Figure 2.7. *Two different chip formation modes: (a) ductile mode, (b) brittle mode. (Source: [CAI 07b] National University of Singapore)*

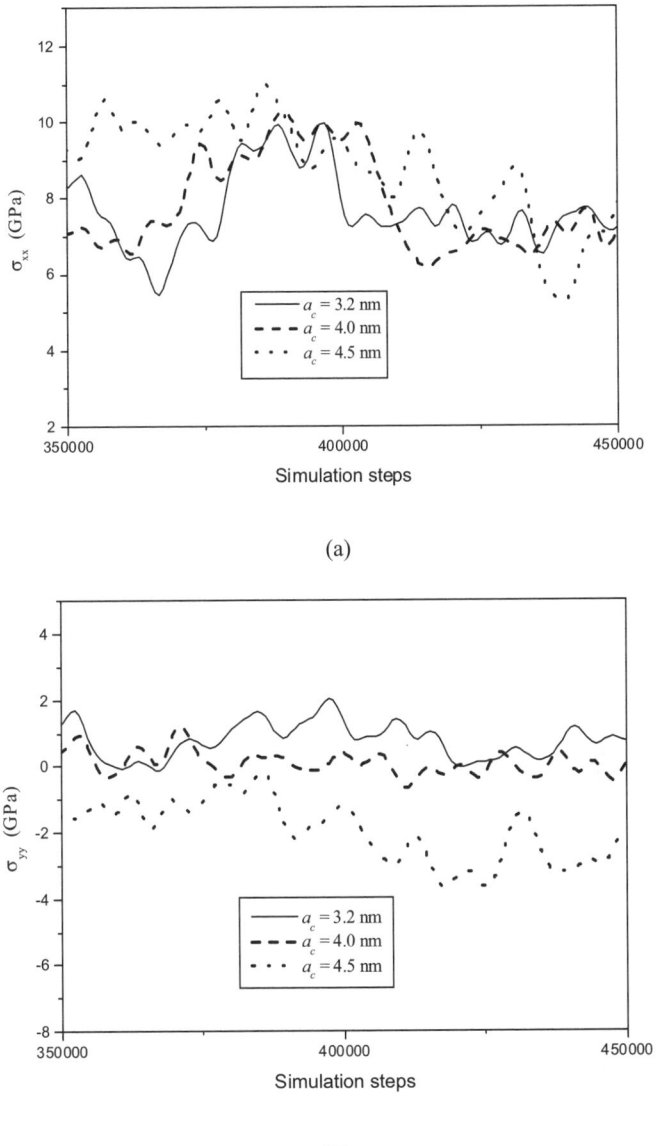

Figure 2.8. *The variations of normal stresses with varying the undeformed chip thickness against the tool cutting edge radius R = 4.0 nm: (a) σ_{xx}, and (b) σ_{yy}. (Source: [CAI 07b] National University of Singapore)*

As shown in Figure 2.7(a), when the undeformed chip thickness was smaller than the tool cutting edge radius, the stress σ_{yy} was compressive and no peak deformation zone occurred in the chip formation zone. Under this condition, the workpiece material next to the chip formation zone was purely compressive stressed. Therefore, there was no crack initiation.

In contrast, as shown in Figure 2.7(b), when the undeformed chip thickness was equal to or larger than the tool cutting edge radius, there was a peak deformation zone generated in the chip formation zone. Unlike ductile materials, which have much larger fracture strains, silicon is a kind of brittle material with extremely small fracture strain. As a result, with the tensile stress in the y direction and compressive stress in the x direction, the workpiece material will be split, initiating a crack in the material next to the peak deformation zone, forming a crack initiation zone next to the peak deformation zone. In this crack initiation zone, the crack would be formed, which propagates along the direction of crack initiation – from the connecting point of the tool rake face and edge arc to the deformation peak of chip formation. This explains the phenomena that in nanoscale cutting of silicon, when the undeformed chip thickness is larger than the tool cutting edge radius, crack initiation and propagation occur and the chip formation is in a brittle mode.

2.4. Machined surfaces in relation to chip formation mode

Experimental cutting tests have been conducted to study the machined surfaces in relation to the chip formation mode in cutting of silicon wafer. Figure 2.9(a) shows the ultra-precision lathe (Toshiba ULG-100) used for nanoscale cutting. Figure 2.9(b) shows the setup for the nanoscale cutting experiments.

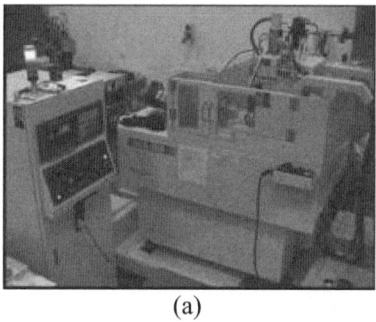
(a)

(b)

Figure 2.9. *(a) Ultra-precision lathe (feedrate resolution up to 1 nm) used in the experiments; (b) Experimental setup for the nanoscale ductile mode cutting of silicon wafer with a single crystal diamond tool. (Source: [LIU 07] National University of Singapore)*

The cutting conditions were controlled such that the undeformed chip thickness was smaller than the tool cutting edge radius for all the cutting tests. The results showed that under the conditions, the machined silicon workpiece surfaces were free of fractures (see Figure 2.10), except for the case of a tool cutting edge radius larger than 800 nm, in which a fractured surface was generated even though the undeformed chip thickness was smaller then the cutting edge radius, indicating an upper bound of the cutting edge radius for ductile mode cutting, which is about 800 nm for cutting silicon and therefore nanoscale is required for ductile mode cutting of silicon.

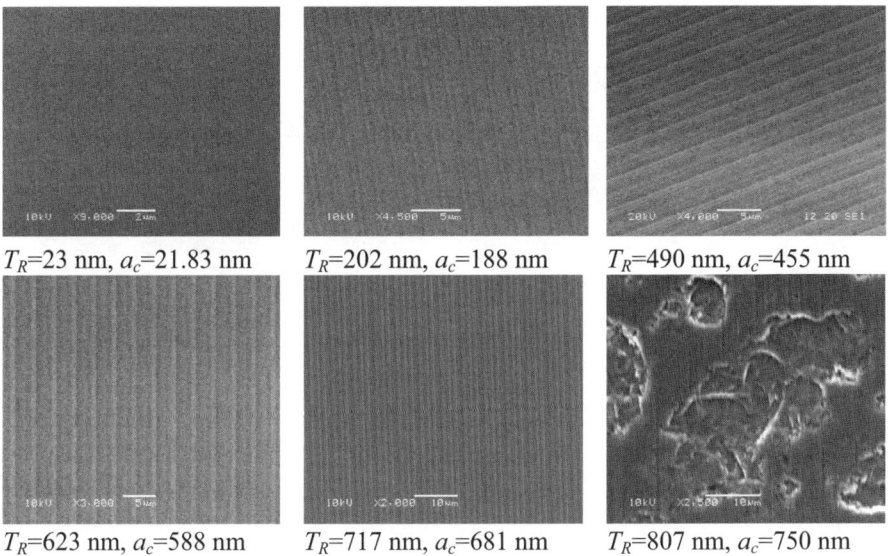

T_R=23 nm, a_c=21.83 nm T_R=202 nm, a_c=188 nm T_R=490 nm, a_c=455 nm

T_R=623 nm, a_c=588 nm T_R=717 nm, a_c=681 nm T_R=807 nm, a_c=750 nm

Figure 2.10. *The machined silicon workpiece surfaces from cutting using a range of tool edge radius with the undeformed chip thickness kept smaller than the tool edge radius, showing an upper bound value of 717 nm for the tool edge radius for ductile mode chip formation that generates a fracture-free surface. (Source: [ARE 05] National University of Singapore)*

The study of the machined surface was further carried out with investigations into the subsurface damage. It was found that when the machined surface was free of fractures as a result of ductile chip formation, the subsurface was also free of fractures, as shown in Figure 2.11(a) and (b). However, when the chip formation was in a brittle mode and the machined surface was fractured, the subsurface was also fractured, as shown in Figure 2.11(c) and (b).

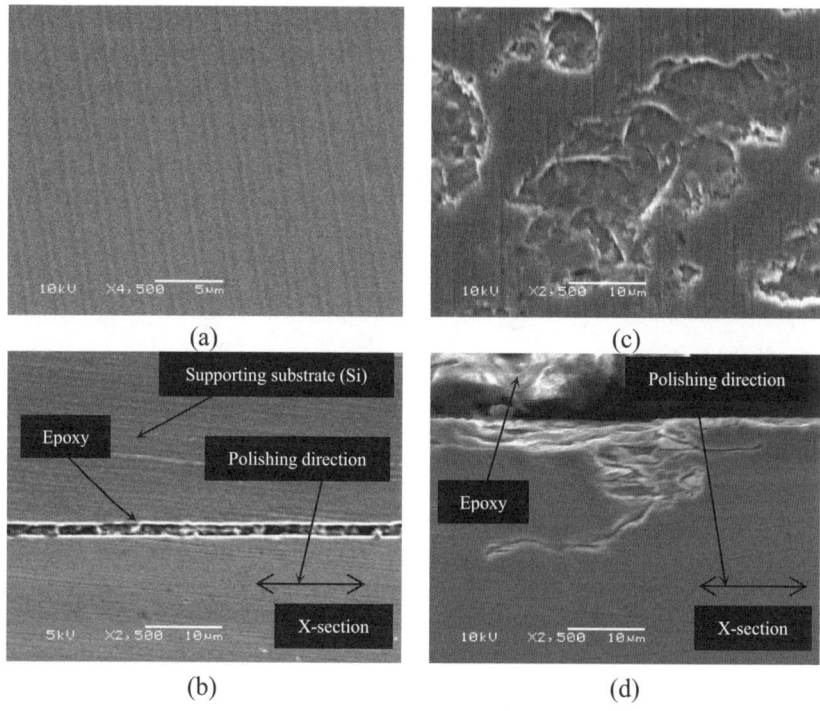

Figure 2.11. *SEM photographs of machined surface and subsurface for cutting with the tool edge radius of 202 nm and undeformed chip thickness of 188 nm: (a) machined surface, and (b) x-section; SEM photographs of machined surface and subsurface for cutting with the tool edge radius of 807 nm and undeformed chip thickness of 750 nm: (c) machined surface, and (d) x-section. (Source: [ARE 05] National University of Singapore)*

2.5. References

[ARE 05] AREFIN S., LI X. P., RAHMAN M. HE T., "Machined surface and subsurface in relation to cutting edge radius in nanoscale ductile cutting of silicon", *Transactions of the North American Manufacturing Research, Institution of SME*, Vol. 33, 2005, pp. 113-119.

[ARE 07] AREFIN, S, LI X. P., RAHMAN M., LIU K., "The upper bound of tool edge radius for nanoscale ductile mode cutting of silicon wafer", *Journal of Advanced Manufacturing Technology*, Vol. 31, 2007, pp. 655-662.

[BLA 90] BLAKE, P.N., SCATTERGOOD, R.O., "Ductile-regime machining of germanium and silicon", *Journal of American Ceramic Society*, Vol. 73, 1990, pp. 949-957.

[CAI 07a] CAI, M.B, LI X.P., RAHMAN, M., "Study of the mechanism of groove wear of the diamond tool in nanoscale ductile mode cutting of monocrystalline silicon", *Journal of Manufacturing Science and Engineering, Transactions of the ASME*, Vol. 129, 2007, pp. 281-286.

[CAI 07c] CAI M., LI X., RAHMAN M., "Molecular dynamics modelling and simulation of nanoscale ductile cutting of silicon", *International Journal of Computer Applications in Technology*, Vol. 28, No. 1, 2007, pp.2–8.

[CAI 07b] CAI M., LI X., RAHMAN M., TAY A.A.O., "Crack initiation in relation to the tool edge radius and cutting conditions in nanoscale cutting of silicon", *International Journal of Machine Tools and Manufacture*, Vol. 47, 2007, pp.562-569.

[CHE 00] CHEONG, W.C.D., ZHANG, L.C. "Molecular dynamics simulation of phase transformation in silicon monocrystals due to nano-indentation", *Nanotechnology*, Vol. 11, 2000, pp. 173–180.

[FAN 98] FANG, F.Z., VENKATESH, V.C., "Diamond cutting of silicon with nanometric finish", *Annals of the CIRP*, Vol. 47, 1998, pp. 45-49.

[GAO 03] GAO, F., HE, J., WU, E., LIU, S., YU, D., LIU, D., ZHANG, S., TIAN, Y. "Hardness of covalent crystals", *Physical Review Letters*, Vol. 91, 2003, p. 501-502.

[GLA 81] GLARDON, R.E., FINNE, I., "Some observations on the wear of single point diamond tools used for machining glass", *J. Mater. Sci.*, Vol. 16. 1981, pp. 1776-1784.

[INA 97] INAMURA, T., SHIMADA, S., NAKAHARA, N., "Brittle/ductile transition phenomena observed in computer simulations of machining defect-free monocrystalline silicon", *Annals of the CIRP*, Vol. 46, 1997, pp. 31-34.

[INA 99] INAMURA, T., SHIMADA, S., TAKEZAWA, N., IKAWA, N., "Crack initiation in machining monocrystalline silicon", *Annals of the CIRP*, Vol. 48, 1999, pp. 81-84.

[KIM 97] KIM, J.D., CHOI, I.H., "Micro surface phenomenon of ductile cutting in the ultrasonic vibration cutting of optical plastics", *Journal of Materials Processing Technology*, Vol. 68, 1997, pp. 89-98.

[KIM 98] KIM, J.D., CHOI, I.H., "Characteristics of chip generation by ultrasonic vibration cutting with extremely low cutting velocity", *The International Journal of Advanced Manufacturing Technology*, Vol. 14, 1998, pp. 2-6.

[KOM 01] KOMANDURI, R., CHANDRASEKARAN, N. and RAFF, L.M., "Molecular dynamics simulation of the nanometric cutting of silicon", *Philosophical Magazine B*, Vol. 81, 2001, pp. 1989–2019.

[LEU 98] LEUNG, T.P., LEE, W.B. and LU, X.M., "Diamond turning of silicon substrates in ductile-regime", *Journal of Materials Processing Technology*, Vol. 73, 1998, pp. 42-48.

[LI 03] LI, X P, RAHMAN M., LIU K., NEO K. S., CHAN C. C., "Nanoprecision measurement of diamond tool edge radius for wafer fabrication", *Journal of Materials Processing Technology*, Vol. 140, 2003, pp. 358-362.

[LI 05] LI, X P, HE T. RAHMAN M., "Tool wear characteristics and their effects on nanoscale ductile mode cutting of silicon wafer", *Wear*, Vol. 259, 2005, pp. 1207-1214.

[LI 07] LI, X P, AREFIN S., CAI M. B., RAHMAN M., LIU K., TAY A. A. O., "Effect of Cutting Edge Radius on Machined Surface in Nanoscale Ductile Mode Cutting of Silicon Wafer", *Proceedings of the Institution of Mechanical Engineers Part B Journal of Engineering Manufacture*, Vol. 221, 2007, pp. 213-220.

[LIU 04] LIU K., LI X.P., RAHMAN M., LIU X.D., "A study of the cutting modes in the grooving of tungsten carbide", *International Journal of Advance Manufacturing Technology*, Vol. 24, 2004, pp. 321–326.

[LIU 07] LIU, K, LI X. P., RAHMAN M., NEO K. S., LIU X. D., "A study of the effect of tool cutting edge radius on ductile cutting of silicon wafers", *International Journal of Advanced Manufacturing Technology*, Vol. 32, 2007, pp. 631-637.

[LIU 08] LIU, K, LI X. P., RAHMAN M., "Characteristics of ultrasonic vibration assisted ductile mode cutting of tungsten carbide" *Journal of Advanced Manufacturing Technology*, Vol. 35, 2008, pp. 833-841.

[MOR 92] MORIWAKI, T., SHAMOTO, E., INOUE, K., "Ultraprecision ductile cutting of glass by applying ultrasonic vibration", *Annals of the CIRP*, Vol. 41, 1992, pp. 141-144.

[PUT 95] PUTTICK, K.E., WHITMORE, L.C., ZHDAN, P., GEE, A.E., CHAO, C.L. "Energy scaling transitions in machining of silicon by diamond", *Tribology International*, Vol. 28, 1995, pp. 349-355.

[TOH 86] TOH, S.B., MCPHERSON, R., "Fine scale abrasive wear of ceramics by a plastic cutting process", in *Science of Hard Materials*, ed. E.A. Almond, C.A. Brookes and R. Warren, 1986, pp. 865-871, Bristol and Boston: Adam Hilger Ltd, Rhodes.

[TON 90] TONSHOFF H.K., SCHMIEDEN W.V., INASAKI I., KONIG W., SPUR, G.. "Abrasive machining of silicon", in *Annals of the CIRP* 39 (2), 1990, pp. 621-635.

[YAN 84] YANG K.H., "An etch for delineation of defects in silicon", *Journal of the Electrochemical Society: Solid-State Science and Technology*, Vol. 131(5), 1984, pp. 1140-1145.

[YAN 01] YAN J., YOSHINO M., KURIYAGAWA T., SHIRAKASHI T., SYOJI K., KOMANDURI R., "On the ductile machining of silicon for micro electro-mechanical systems (MEMS), opto-electronic and optical applications", *Mater. Sci. Eng. A*, Vol. 297, 2001, pp. 230–234.

[YAN 03] YAN J.W., SYOJI K., TAMAKI J., "Some observations on the wear of diamond tools in ultra-precision cutting of single-crystal silicon", *Wear*, Vol. 255, 2003, pp. 1380-1387.

[ZHA 98] ZHANG L.C., TANAKA H., "Atomic scale deformation in silicon monocrystals induced by two-body and three-body contact sliding", *Tribology International*, Vol. 31, 1998, pp. 425-433.

[ZHA 99] ZHANG L.C., TANAKA H., "On the mechanics and physics in the nano-indentation of silicon monocrystals", *JSME International Journal A*, Vol. 42, 1999, pp. 546-559.

Chapter 3

Diamond Tools in Micromachining

3.1. Introduction

Diamond tools are used extensively in micromachining processes. There are many coating processes ranging from traditional electroplating to the more advanced laser or ion-assisted deposition. However, the choice of deposition technology depends on many factors including substrate properties, component dimensions and geometry, production requirements, and the coating specification needed for the application of interest. For complex geometric components, small feature sizes, good reproducibility and high product throughput chemical vapor deposition (CVD) is a highly effective technology. For example, low-pressure and plasma assisted CVD is a well-established technology for semiconductor devices that possess very small feature sizes and complex geometric arrangements. This chapter explains how diamond tools are created and how they are used in micromachining processes.

3.2. Diamond technology

The reactor system (comprising the reaction chamber and all associated equipment) for carrying out CVD processes must provide several basic functions common to all types of systems It must allow transport of the reactant and diluent gases to the reaction site, provide activation energy to the reactants (heat, radiation, plasma), maintain a specific system pressure and temperature, allow the chemical processes for film deposition to proceed optimally, and remove the by-product gases

---

Chapter written by Waqar AHMED, Mark J. JACKSON and Michael D. WHITFIELD.

46 Nano and Micromachining

and vapors. These functions must be implemented with adequate control, maximum effectiveness and complete safety [1-5].

Chemical vapor deposition is a crystal growth process used not only for diamond but also for a range of different semiconductors and other crystalline materials, such as silicon or gallium arsenide. These industrial fields are diverse and range from gas turbines to gas cookers and from coinage to nuclear power plants [6,7].

The CVD process relies first on species generation, produced by the reaction of the element that is to be deposited with another element that results in a substantial increase of the depositing elements vapor pressure. Second, this volatile species is then passed over or allowed to come into contact with the substrate being coated. This substrate is held at an elevated temperature, typically from 800-1,150°C. Finally, the deposition reaction usually occurs in the presence of a reducing atmosphere, such as hydrogen. The film properties can be controlled and modified by varying the problem parameters associated with the substrate, the reactor and gas composition.

3.2.1. *Hot Filament CVD (HFCVD)*

In the early 1970s it was suggested that the simultaneous production of atomic hydrogen during hydrocarbon pyrolysis may enhance the deposition of diamond. Soviet researchers who generated H by dissociating H_2 using an electric discharge or a hot filament tested this suggestion [8]. It was observed that atomic hydrogen could easily be produced by the passage of H_2 over a refractory metal filament, such as tungsten, heated to temperatures between 2,000 and 2,500 K. When atomic hydrogen was added to the hydrocarbon, typically with a C/H ratio of ~0.01, it was observed that diamond could be deposited while graphite formation was suppressed. The generation of atomic hydrogen during diamond CVD enabled (a) a dramatic increase in the diamond deposition rate to approximately 1 μm hr^{-1} and (b) the nucleation and growth of diamond on non-diamond substrates [8-13]. Due to its inherent simplicity and comparatively low operating cost, HFCVD has become very popular in industry. Table 3.1 outlines typical deposition parameters used when growing diamond films by this technique.

A wide variety of refractory materials have been used as filaments including tungsten, tantalum, and rhenium due to their high electron emissivity. Refractory metals that form carbides (e.g., tungsten and tantalum) typically must carburize their surface before supporting the deposition of diamond films. The process of filament carburization results in the consumption of carbon from the CH_4, and thus a specific incubation time is needed for the nucleation of diamond films. Therefore, this process may affect the early stages of film growth, although it is insignificant over longer periods. Furthermore, the volume expansion due to carbon incorporation leads to cracks along the length of the wire. The development of these cracks is undesirable, as it reduces the lifetime of the filament.

Gas mixture	Total pressure (Torr)	Substrate temperature (K)	Filament temperature (K)
CH_4 (0.5-2.0%)/H_2	10-50	1,000-1,400	2,200–2,500

Table 3.1. *Typical deposition parameters used when growing diamond films by HFCVD*

It is believed that thermodynamic near-equilibrium is established in the gas phase at the filament surface. At temperatures around 2,300 K, molecular hydrogen dissociates into atomic hydrogen and methane transforms into methyl radicals, acetylene species, and other hydrocarbons that are stable at these elevated temperatures. Atomic hydrogen and the high-temperature hydrocarbons then diffuse from the filament to the substrate surface. Although the gaseous species generated at the filament are in equilibrium at the filament temperature, the species are at a super equilibrium concentration when they arrive at the much cooler substrate. The reactions that generate these high-temperature species (e.g. C_2H_2) at the surface of the filament or anywhere where there are hydrogen atoms proceed faster than any reactions that decompose these species during the transit time from the filament to the substrate. Consider the equilibrium between methane and acetylene:

$$2CH_4 \rightleftharpoons C_2H_2 + 3H_2$$

At the filament surface, the reaction is immediately driven to the right, creating acetylene. After acetylene diffuses to the substrate, thermodynamic equilibrium at a substrate temperature of ~1,100 K induces the formation of methane, but the reverse reaction proceeds much slower. Solid carbon precipitates on the substrate in order to reduce the equilibrium concentration of species such as acetylene in the gas phase. The diamond allotrope of carbon is "stabilized" by a concurrent super equilibrium concentration of atomic hydrogen. This simple explanation emphasizes the importance of reaction kinetics in diamond synthesis by HFCVD. The reader is directed to references [9-43] for a full description of the nature of CVD technology for creating diamond tools and diamond surfaces.

3.3. Preparation of substrate

3.3.1. *Selection of substrate material*

Deposition of adherent high-quality diamond films onto substrates such as cemented carbides, stainless steel and various metal alloys containing a transition element has proved to be problematic. In general, the adhesion of the diamond films to the substrate is poor and the nucleation density is very low [44-51]. Mainly refractory materials such as W (WC-Co), Mo and Si have been used as substrate materials. Materials that form carbide are found to support diamond growth. However, materials such as Fe and steel possess a high mutual solubility with carbon, and only graphitic deposits or iron carbide result during CVD growth on these materials. For applications in which the substrate needs to remain attached to the CVD diamond film, it is necessary to choose a substrate that has a similar thermal expansion coefficient to that of diamond. If this is not done, the stress caused by the different rates of contraction on cooling after deposition will cause the film to delaminate from the substrate. The influence of different metallic substrates on the diamond deposition process has been examined.

Interactions between substrate materials and carbon species in the gas phase are found to be particularly important and lead to either carbide formation or carbon dissolution. Carbides are formed in the presence of carbon-containing gases on metals such as molybdenum, tungsten, niobium, hafnium, tantalum and titanium. The carbide layer formed allows diamond to form on it since the minimum carbon surface concentration required for diamond nucleation cannot be reached on pure metals. As the carbide layer increases in thickness, the carbon transport rate to the substrate decreases until a critical level is reached where diamond is formed [52-58]. Substrates made from metals of the first transition group such as iron, cobalt, and nickel, are characterized by high dissolution and diffusion rates of carbon into those substrates (Table 3.2) [59]. Owing to the absence of a stable carbide layer, the incubation time required to form diamond is higher and depends on substrate thickness. In addition, these metals catalyze the formation of graphitic phases, which is reflected in the graphite-diamond ratio during the deposition process, yielding a low quality diamond. The importance of this mechanism in relation to diamond deposition decreases from iron to nickel, corresponding to a gradual filling of the 3D-orbital [59]. This effect occurs whenever the metal atoms come into contact with the carbon species, which can take place on the substrate or in the gas phase [60].

3.3.2. Pre-treatment of substrate

In order for continuous film growth to occur, a sufficient density of crystallites must be formed during the early stages of growth. In general, the substrate must undergo a nucleation enhancing pre-treatment to allow this.

	α-Fe	γ-Fe	Co	Ni
Solubility of carbon (wt.%)	1.3	1.3	0.1	0.2
Carbon diffusion rate (cm/s)	2.35×10^{-6}	1.75×10^{-8}	2.46×10^{-8}	1.4×10^{-8}

Table 3.2. *Solubility and diffusion rates of carbon atoms in different metals at 900°C*

This is particularly true for Si wafer substrates that have been specially polished to be smooth enough for micro-electronic applications. Substrates may be pre-treated by a variety of methods including:

– Abrasion with small (~nm/μm size) hard grits (e.g. diamond, silicon carbide).

– Ultrasonication of samples in slurry of hard grit (e.g. diamond).

– Chemical treatment (acid etching and Murakami agent).

– Bias enhanced nucleation (BEN) (negative/positive substrate biasing).

– Deposition of hydrocarbon/oil coatings.

The basis for most of these methods is to produce scratches, which provide many sites for diamond nucleation of crystallites. It is also possible that small (~nm size) particles of diamond, produced during abrasion with diamond grit, become embedded in the substrate, and that CVD diamond grows on this material [61].

It could be desirable to produce nucleation sites without damage to the underlying substrate. This is particularly important for some applications such as diamond electronics and optical components. One method for encouraging nucleation without damaging the substrate material has been developed: bias enhanced nucleation (BEN).

(a) Pre-treatment on Mo/Si substrate

Prior to pre-treatment the Si/Mo substrate is ultrasonically cleaned in acetone for 10 minutes to remove any unwanted residue on the surface. Abrasion with 1 µm sizes of diamond powder is performed for 5 minutes. Alternatively, the substrate is immersed in diamond solution containing 1-3 µm of diamond particles and water for 1 hr in an ultrasonic bath. These methods produce scratches on the surface, which create many nucleation sites. The substrates are then washed with acetone in the ultrasonic bath for 10 minutes. SEM and energy-dispersive x-ray spectroscopy (EDX) characterized the abraded surface of substrates.

(b) Pre-treatment on WC-Co substrate

The application of diamond coatings on cemented tungsten carbide (WC-Co) tools has attracted much attention in recent years in an effort to improve cutting performance and tool life. However, deposition of adherent high-quality diamond films onto substrates such as cemented carbides, stainless steel and various metal alloys containing transition element has proved to be problematic. In general, the adhesion of the diamond films to the substrates is poor and the nucleation density is very low [44-50]. WC-Co tools contain 6% Co and 94% WC substrate with grain size 1-3 micron is desirable for diamond coatings.

In order to improve the adhesion between diamond and WC substrates it is necessary to etch away the surface Co and prepare the surface for subsequent diamond growth. This particularly applies to the Co binder, which provides additional toughness to the tool, but is hostile to diamond adhesion. The adhesion strength of diamond films is relatively poor, and can lead to catastrophic coating failure in metal cutting [59]. The Co binder can also suppress diamond growth, favoring the formation of non-diamond carbon phases and resulting in poor adhesion between the diamond coating and the substrate [62]. Most importantly, it is difficult to deposit adherent diamond onto untreated WC-Co substrates.

Poor adhesion can be related to the cobalt binder, which is present to increase the toughness of the tool; however, it suppresses diamond nucleation and causes deterioration of diamond film adhesion. To eliminate this problem, it is usual to pre-treat the WC-Co surface prior to CVD diamond deposition. Various approaches have been used to suppress the influence of Co and to improve adhesion. Therefore, a substrate pre-treatment, for reducing the surface Co concentration and achieving a proper interface roughness, will enhance the surface readily available for coating process [62]. For example, chemical treatment using a Murakami agent and acid etching has been used successfully for removal of the Co binder from the substrate surface [63].

The WC-Co substrates (flat) used were 10 x 10 mm in length and 3 mm in thickness. The hard metal substrates used were WC-6wt % Co with WC average grain size of 0.5 μm (fine grain) and 6 μm (coarse grain). Table 3.3 shows that the data for a substrate, which consists of the chemical composition, density, and hardness of samples, used for diamond deposition.

The Co cemented tungsten carbide (WC-Co) rotary tools (microtools), 20 mm in length, including the microtool head (WC-Co) and shaft (Fe/Cr) and ~1 mm in diameter, were also used. Prior to pre-treatment both sets of substrates are ultrasonically cleaned in acetone for 10 minutes to remove any loose residues. The following two-step chemical pre-treatment procedure is used. A first step etching, using Murakami's reagent (10 g $K_3Fe(CN)_6$ + 10 g KOH + 100 ml water), is carried out for 10 minutes in an ultrasonic bath to etch the WC substrate, followed by a rinse with distilled water. The second step etching is performed using an acid solution of hydrogen peroxide (3 ml (96% wt.) H_2SO_4+ 88 ml (30% w/v) H_2O_2), for 10 s, to remove Co from the surface. The substrates are then washed again with distilled water in an ultrasonic bath. After wet treatment the microtool is abraded with synthetic diamond powder (1 μm grain size) for 5 minutes, followed by ultrasonic treatment with acetone for 20 minutes. The etched surface of substrates can be characterized by SEM and energy-dispersive x-ray spectroscopy (EDX).

WC grain size (μm)	WC	Co	TaC	Density (g/cm^3)	Hardness (HRA)
WC fine grain 0.5 μm	94.2	5.8	0.2	14.92	93.40
WC coarse grain 6 μm	94.0	6.0		14.95	88.50

Table 3.3. *WC-Co insert chemical composition*

3.4. Modified HFCVD process

3.4.1. *Modification of filament assembly*

The filament material and its geometric arrangement are important factors to consider in order to have improved coatings using the CVD method. Therefore, in order to optimize both the filament wire diameter and the filament assembly/geometry, it is necessary to understand the temperature distributions of the filament. Research by the author indicated that the best thermal distribution and diamond growth uniformity is obtained using tantalum wires of 0.5 mm in diameter. To ensure uniform coating around the cylindrical shape samples (microtools), tools were positioned centrally and coaxially within the filament coils, the six-spiral (coil) filament being made with 1.5 mm spacing between the coils (Figure 3.1a and b).

Tantalum wire of 0.5 mm in diameter and 12-14 cm in length is used as the hot filament. The filament is mounted vertically with the microtool held in between the filament coils, as opposed to the horizontal position used in the conventional HFCVD system. To ensure uniform coating the microtool is positioned centrally and coaxially within the filament coils. A schematic diagram of the modified HFCVD system is presented in Figure. 3.1 and has been designed for a microtool or a wire with a similar diameter. The new vertical filament arrangement used in the modified HFCVD system enhances the thermal distribution, ensuring uniform coating, increased growth rates and higher nucleation densities.

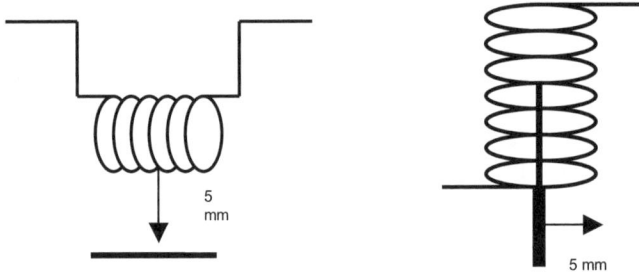

Figure 3.1. *(a) Conventional filament arrangement, (b) modified filament arrangement*

3.4.2. Process conditions

The CVD reactor is a cylindrical stainless steel chamber measuring 20 cm in diameter and 30 cm in length. Diamond films were deposited onto the tool cutting edges at 5 mm distances from the filament. The gas source used during the deposition process is composed of a mixture containing 1% methane with an excess of hydrogen; the volume flow rate for hydrogen is 100 sccm, while the volume flow rate for methane is 1 sccm. The deposition time and pressure in the vacuum chamber were 5-15 hours and 20 Torr (2.66 k Pa) respectively. The substrate temperature was measured by a K-type thermocouple mounted on a molybdenum substrate holder.

The depositions are carried out between 800 and 1,000°C. The filament temperature is measured using an optical pyrometer and found to be between 1,800 and 2,100°C depending on the filament position. A summary of the process conditions is shown in Table 3.4.

3.5. Nucleation and diamond growth

The growth of thin diamond films at low pressures, at which diamond is metastable, is one of the most exciting developments in materials science of the last two decades. However, low growth rates and poor quality currently limit applications. Diamond growth is achieved by a variety of processes using very different means of gas activation and transport. Generalized models coupled with experiments show how process variables, such as gas activation temperature, pressure, characteristic diffusion length and source gas composition, influence diamond growth rates and diamond quality. The modeling is sufficiently general to permit comparison between growth methods. The models indicate that typical processes, e.g., hot-filament, microwave and thermal plasma reactors, operate at pressures where concentrations of atomic hydrogen [H] and methyl radicals [CH_3] reach their peak. The results strongly suggest that the growth rate peaks with pressure because of changes in the gas phase concentrations rather than changes in substrate temperature.

Process variables	Operating parameters
Tantalum filament diameter (mm)	0.5
Deposition time (hours)	5-15
Gas mixture	1% CH_4 in excess H_2
Gas pressure (Torr)	20 (2.66 k Pa)
Substrate temperature (°C)	800-1000
Filament temperature (°C)	1800-2100
Substrate (WC-Co/Mo/Ti) diameter (mm)	Wire/drill/microtool (approx. 1 mm)
Distance between filament and substrate (mm)	5
Pre-treatment (Murakami etching and acid etching)	20 minutes plus 10 seconds

Table 3.4. *Process conditions used for diamond film deposition on microtools*

The results also suggest that, at an atmospheric pressure of 1, using only hydrocarbon chemistry, growth rates saturate at gas activation temperatures above 4,000 K. Models of defect incorporation indicate that the amount of sp^2, non-diamond material incorporated in the diamond, is proportional to [CH_3]/[H] and therefore can be correlated with the controllable process parameters. The unusual and interesting connection between diamond nucleation and growth using the process of vapor synthesis is essentially quite simple. Carbon containing precursor molecules (like CH_4) are excited and/or dissociated and subsequently condense via a free dangling bond of the radical in diamond configuration on a surface [64-65] (Figure. 3.2). A nucleation pathway occurs through a stepwise process including the formation of extrinsic (pre-treatment) or intrinsic (*in situ*) nucleation sites, followed by the formation of carbon-based precursors. It is believed that nucleation sites could be either grooves of scratching lines or protrusions produced by etching redeposition.

The gas activation is performed either by hot filaments, microwaves, or radio frequency plasmas. The most crucial parameter in all this processes is, besides a carbon source, the presence of large amounts of atomic hydrogen. The role of atomic hydrogen in the process is:

– creation of active growth sites on the surface;

– creation of reactive growth species in the gas-phase;

– etching of non-diamond carbon (like graphite) graphitic, sp^2, precursors will be explored.

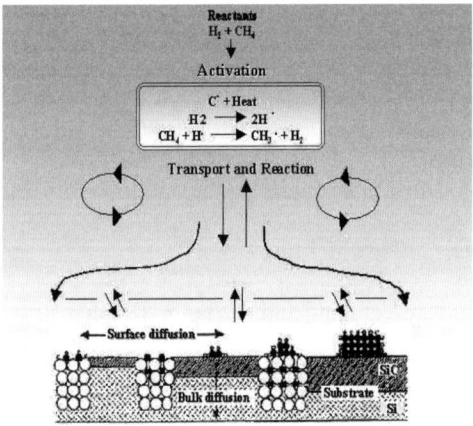

Figure 3.2. *Schematic diagram of diamond nucleation and growth*

3.5.1. *Nucleation*

Diamond nucleation is a critical and necessary step in the growth of thin diamond films, because it strongly influences diamond growth, film quality and morphology [66]. Diamond growth begins when individual carbon atoms nucleate onto the surface to initiate the beginnings of an sp³ tetrahedral lattice. There are two types of diamond growth:

– *Homoepitaxial growth:* this is an application of diamond substrates. The template for the required tetrahedral structure is already there, and the diamond lattice is just extended atom-by-atom as deposition proceeds.

– *Heteroepitaxial growth: this* uses the non-diamond substrates. There is no such template for the C atoms to follow, and those C atoms that deposit in non-diamond forms are immediately etched back to the gas phase by reaction with atomic [H].

To deal with the problem of the initial induction period before diamond starts to grow, the substrate surface often undergoes pre-treatment prior to deposition in order to reduce the induction time for nucleation and to increase the density of nucleation sites. There are two main methods to apply this pre-treatment.

Generally, seeding or manual abrasion with diamond powder, or immersion in diamond paste containing small crystallites processed in an ultrasonic bath enhances nucleation. The major consideration is the nucleation mechanism of diamond on non-diamond substrates. It has been shown that the pre-abrasion of non-diamond substrates reduces the induction time for nucleation by increasing the density of nucleation sites. The abrasion process can be carried out by mechanically polishing the substrate with abrasive grit, usually diamond power of 0.1-10 µm particle size, although other nucleation methods do exist such as bias-enhanced nucleation, which is used in heteroepitaxial growth. The most promising *in situ* method for diamond nucleation enhancement is substrate biasing. In recent years, more controlled techniques, such as bias-enhanced nucleation and nanoparticle seeding, have been used to deposit smoother films [67-68]. In this method, the substrate is biased negatively during the initial deposition stage [69]. Before CVD diamond deposition the filament was pre-carburized for 30 minutes in 3% methane with excess hydrogen to enhance the formation of a tantalum carbide layer on the filament surface in order to reduce the tantalum evaporation during diamond deposition [70].

3.5.2. *Bias-enhanced nucleation (BEN)*

The substrate can be biased both negatively and positively; however, there is much research work and a large volume of literature on negative biasing. Negative substrate biasing is attractive because it can be controlled precisely; it is carried out *in situ*, gives good homogeneity, and results in improved adhesion. On flat substrates, such as copper and silicon, biasing has been shown to give better adhesion, improved crystallinity and smooth surfaces.

A negative bias voltage up to –300 V has been applied to the substrate relative to the filament. This produced emission currents up to 200 mA. The nucleation times used were between 10 and 30 minutes. In the activated deposition chamber, CH_4 and H_2 were decomposed into various chemical radicals species CH_3, C_2H_2, CH_2, CH, C and atomic hydrogen H by the hot tantalum filament. Methyl radicals and atomic hydrogen are known to play important roles in diamond growth. In the biasing process electrons were emitted from the diamond-coated molybdenum substrate holder and moved to the filament after they gained energy from the electrical field. When the negative bias was applied to the anode, the voltage was gradually increased until a stable emission current was established and a luminous glow discharge was formed near the substrate [71]. As bias time is increased, the nucleation density also increases. The highest nucleation density was calculated to be 0.9×10^{10} cm^{-2} for a bias time of 30 minutes. At a bias time of 10 minutes the nucleation density obtained was 2.7×10^8 cm^{-2}.

Wang et al. [72] also reported that an increase in the emission current produced higher nucleation densities [73]. Since the biased voltage and emission current are related, the enhancement of the nucleation density cannot be attributed solely to ion bombardment or electron emission of the diamond-coated molybdenum substrate holder, but may be a combination of these mechanisms [74]. Results were based on negative BEN related to the grounded filament. However, it was reported that very low electric biasing current values (μA) were detected for applied substrate biased voltages (either positive or negative). Furthermore, increasing negative biases of up to –200 V resulted in a value of nucleation density similar to that obtained with positively bias-enhanced nucleation related to the filament. In contrast, an application of negative bias to the substrate at –250 V resulted in (10^{10} cm^{-2}) maximum values of nucleation density. The enhancement in the nucleation density can be attributed to the electron current from the filament increasing the decomposition of H_2 and CH_4. The increase in the nucleation density is expected, since negatively biasing the substrate increases the rate of ion bombardment into the surface creating greater numbers and density of nucleation sites. Therefore, the greater the density of nucleation sites, the higher the nucleation density. Kamiya et al. reported that reproducibility of the experiment was poor and that no definite trend in the nucleation density could be found with respect to different bias conditions [74].

3.5.3. Influence of temperature

Temperature is a major factor in influencing the deposition rate, crystallite size and controlling the surface roughness. Variations in the average crystallite diamond size along the length of the substrate (microtool) can be attributed to the difference in the substrate temperature. The substrate temperature from the end to the center of the filament is more accentuated for molybdenum wire with a smaller diameter [9]. Most physical vapor deposition processes operate at conditions where the rate-determining step of the deposition process is the diffusion of precursor gases to the substrate surface. Generally, this results in poor film uniformity in grooves and at the sharp edges. By operating under a kinetic control regime, film uniformity is greatly enhanced.

Deposition temperature can also influence the diamond film thickness in terms of substrate and filament position. Analysis of temperature distribution along the coiled filament showed that the temperature is highest at the center of the filament, with a rapid decrease toward the edges. Generally, higher substrate temperatures increase diamond film growth rate and the crystallite size. At the bottom of the filament coil temperature is lower; therefore, the part of the microtool parallel to the coil at this temperature will be coated with the diamond film at a lower growth rate. Thus, it can be expected that at these regions the film will be thinner. The thermal gradient gives variations in the film thickness and crystal sizes. Generally, with columnar growth, the average crystallite size increases as the films become thicker.

Analysis of temperature distribution along the coiled filament showed that the temperature is highest at the center of the filament with a rapid decrease toward the edges. The microtool substrate and filament temperature have been measured parallel to the positions A, B, and C respectively. This suggests that position A is the hottest, followed by position B and C on the microtool. Variations in the film thickness and crystal sizes are mainly due to thermal gradients at various positions on the microtool. It was found that the coating is thicker at the cutting teeth, with average thickness of about 43 μm due to the slightly higher temperature at the microtool tip because cutting teeth is closer to the filament coil. At the base of the microtool the heat is carried away faster and is therefore at a lower temperature, giving rise to lower growth rates and hence thinner films, at about 23 μm in thickness. Thicker coating at the tip is expected to give the tool longer life. Further work is required to study the effects of film thickness at the tooth tip and at the base on tool performance and lifetime.

Deposition temperature can also influence the diamond film thickness in terms of substrate and filament position. Analysis of temperature distribution along the coiled filament showed that the temperature is highest at the center of the filament with a rapid decrease toward the edges. This suggests that position A is the hottest, followed by position B and C on the microtool. Generally, higher substrate temperatures increase the diamond film growth rate and the crystallite size. At the bottom of the filament coil the temperature is lower; therefore, the part of the microtool parallel to the coil at this temperature will be coated with the diamond film at a lower growth rate. Consequently, it can be expected that at these regions the film will be thinner. The thermal gradient gives variations in the film thickness and crystal sizes. Generally, with columnar growth, the average crystallite size increases as the films become thicker. The films were thicker and the crystallite size was larger at position A compared to position C.

Analysis of temperature distribution along the coiled filament showed that the temperature is highest at the center of the filament with a rapid decrease toward the edges. The microtool substrate and filament temperature have been measured parallel to the positions A, B, and C, respectively. This suggests that position A is the hottest followed by position B and C on the microtool. Variations in the film thickness and crystal sizes are mainly due to thermal gradients at various positions on the microtool.

3.6. Deposition on complex substrates

3.6.1. *Diamond deposition on metallic (molybdenum) wire*

It is difficult to deposit CVD diamond onto cutting tools, which generally have a 3D shape and possess complex geometry and sharp edges, using a single step growth process [76]. The cylindrical shape wire, which has complex geometry, can be used as a model application for depositing diamond on cutting tools such as microtools. The molybdenum (Mo) wires are deposited with CVD diamond by a modified vertical filament approach. After deposition time of 5 hours is complete films of 5 μm thick CVD diamond were obtained. The film morphology showed that it has good uniformity and high purity of diamond. The Raman spectroscopy confirmed that sp^3 diamond peak at wave number 1,332.6 cm^{-1}.

3.6.2. *Deposition on WC-Co microtools*

Deposition of diamond on wires can be readily extended to a microtool used as a machining tool, and MEMS devices. The uniform and adherent coating are essential in order to obtain an improved performance. Figure 3.3 shows the SEM micrographs and the corresponding EDX spectra of the WC-Co microtool before and after the chemical etching process. Before etching, the EDS spectrum (Figure 3.3(a)) shows the peaks for cobalt (Co), carbon (C) and tungsten (W). High cobalt content inhibits diamond deposition, resulting generally in graphitic phases, which degrade the coating adhesion. The Co diffuses to the surface regions, preventing effective bonding between the substrate surface and the film coating. To improve the coating adhesion of diamond on WC-Co tools, several approaches can be employed. For example, firstly, the use of interlayer material such as chromium can act as a barrier against Co diffusion during diamond CVD.

Secondly, the Co from the tool surface can be etched using either chemical or plasma methods. Thirdly, the Co can be converted into stable intermediate interlayer cobalt compounds. These can act as a barrier to Co diffusion from the substrate during film growth [77]. A Murakami solution followed by H_2SO_4/H_2O_2 etching can be used to chemically remove the cobalt from the microtool surface. The EDX spectrum shows that the Co peak has disappeared after etching. This will help to enhance the coating adhesion. Comparison of the SEM micrograph in Figure 3.3(a) and (b) shows that the surface topography is significantly altered after etching in Murakami and H_2SO_4/H_2O_2 solutions. The etching process makes the surface much rougher, with a significant amount of etch pits, which act as low-energy nucleation sites for diamond crystal growth.

An SEM micrograph of a diamond-coated WC microtool is shown in Figure 3.4. Six cutting edges of the microtool tip were coated with a polycrystalline diamond film using the modified vertical HFCVD method. Analysis of the SEM picture shows that the coating uniformly covered the cutting edges as well as the nearby regions in which the placement of the microtool within the filament coils, ensuring uniform deposition. The diamond crystal structure and morphology are uniform and adherent, as shown in Figures 3.4(a) and (b). It is also shows a close-up view of the diamond-coated region of the microtool in Figure 3.4(c).

Typically the crystallite sizes are of the order of 5-8 µm. The visibly adherent diamond coatings on the WC-Co microtools consist of mainly (111) faceted diamond crystals. The design of the filament and substrate in the reactor offer the possibility of uniformly coating cylindrical substrates of even larger diameters. Raman analysis was performed in order to evaluate the diamond carbon-phase quality and film stress in the deposited films. The Raman spectrum showed a single peak at 1,335 cm^{-1} for the tip of the diamond-coated microtool. The Raman spectrum also gives information about the stress in the diamond coatings. The diamond peak is shifted to a higher wave number of 1,335 cm^{-1} than that of natural diamond peak 1,332 cm^{-1} indicating that stress, which is compressive in nature, exists in the resultant coatings [78].

3.6.3. *Diamond deposition on tungsten carbide (WC-Co) microtool*

Laboratory grade tungsten carbide (WC-Co) microtools are used (AT23 LR) with fine WC grain sizes (1 µm) 20-30 mm in length and 1.0-1.5 mm in diameter (supplied by Metrodent Ltd, UK) that are used for the CVD diamond deposition process.

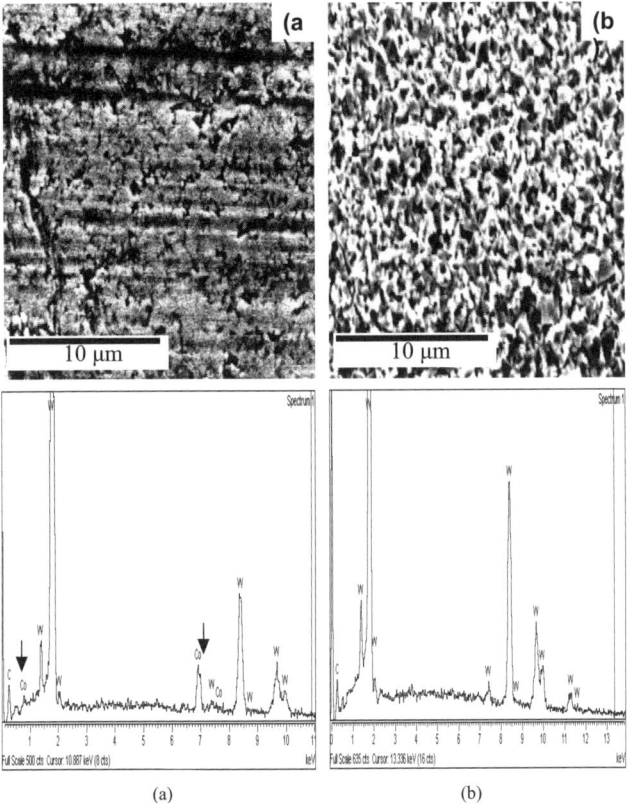

Figure 3.3. *(a) SEM and EDX of WC-Co micro-drill before etching; (b) WC-Co micro-drill after etching*

Diamond Tools in Micromachining 61

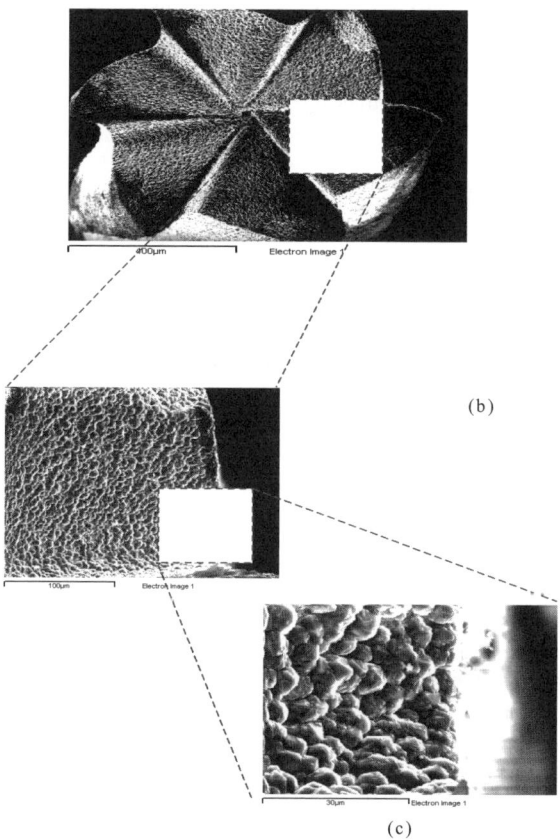

Figure 3.4. *(a) Cutting edge of micro-drill after depositing with CVD diamond; (b) cutting edge of micro-drill uniformly coated with diamond; (c) SEM of micro-drill after coating with diamond*

The WC surface has been etched away with Murakami solution and surface Co has been removed by acid etching followed by ultrasonic washing in distilled water. The EDX results confirmed that there is no indication of Co left on the surface of the etched microtool (Figure 3.5). Diamond films have been deposited onto the cutting edge of the tools at a 5 mm distance from the tantalum wire filament.

Surface morphology of predominantly (111) faceted octahedral shape diamond films was obtained. The film thickness was measured to be 15-17 μm after diamond deposition for 15 hours. Figure 3.6 shows the SEM micrograph of a CVD diamond-coated microtool (AT 23LR) at the cutting edge. The film is homogenous with uniform diamond crystal sizes, typically in the range of 6-10 μm. As expected the surface morphology is rough, making the microtools extremely desirable for abrasive applications.

The Raman spectra shown demonstrate that at the tip, center and end of cutting tool single sharps at 1,336, 1,336 and 1,337 cm^{-1} respectively peaked for different positions along the microtool. The Raman spectrum also gives an indication about the stress in the diamond coating. The diamond peak is shifted to a higher wave number of magnitude (such as 1,336 or 1,337 cm^{-1}) than that normally experienced in an unstressed coating where the natural diamond peak occurs at 1,332 cm^{-1}, thus indicating that the stress is compressive.

3.7. Diamond micromachining

The quality and economy of industrial production processes are to a great extent determined by the selection and the design of appropriate manufacturing operations. For many machining operations, especially for the technologically relevant processing of metallic materials, machining with a geometrically specified cutting edge is applicable. Enhancing the performance of machining operations is therefore an economically important goal.

The cutting tool is the component that is most stressed, and therefore limits the performance in MEMS operations. Thermal and mechanical loading affects the cutting tool edges in a continuous or intermittent way. As a result, in addition to good wear resistance, high thermal stability and high mechanical strength are required for cutting materials. Opposing this objective of an ideal cutting material is the fundamental contradiction of properties hardness, the strength at elevated temperature and wear resistance on the one hand, and bending strength and bending elasticity on the other hand.

Diamond Tools in Micromachining 63

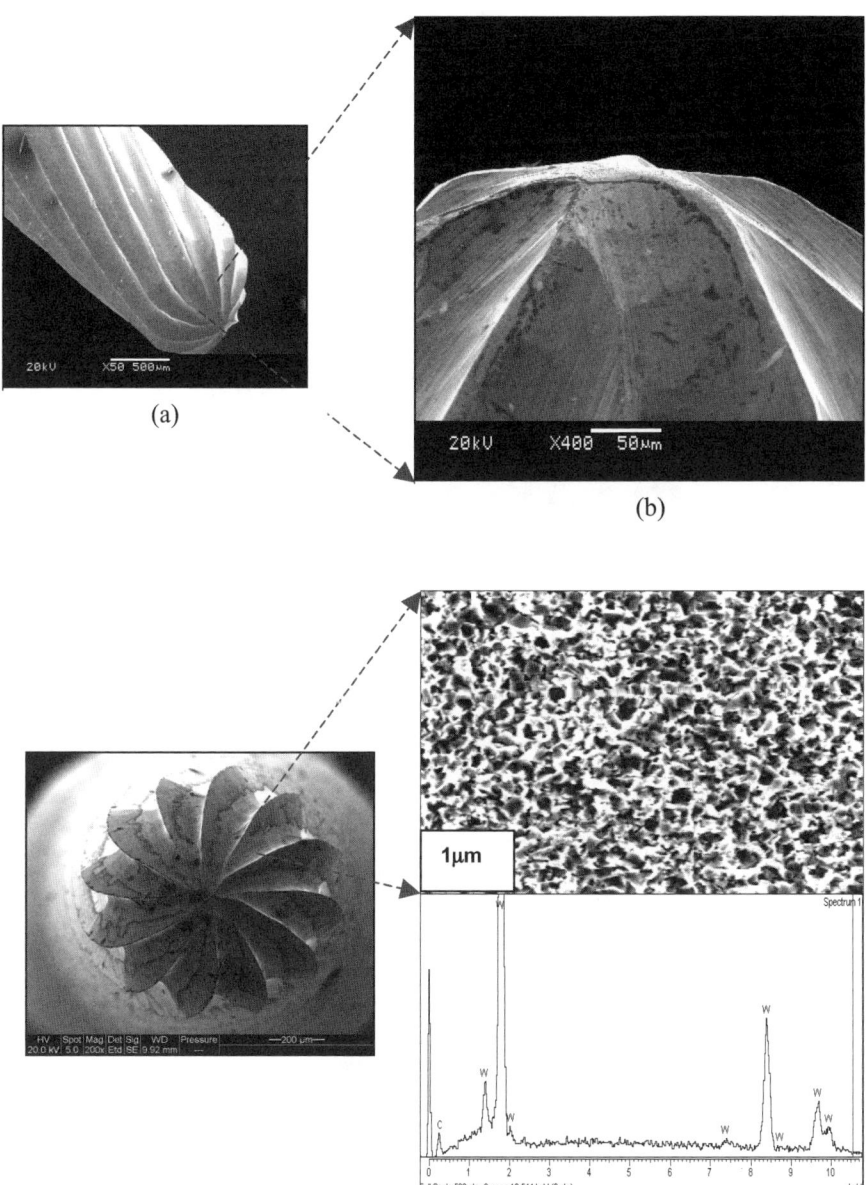

Figure 3.5. *Etched microtool surface after chemical treatment*

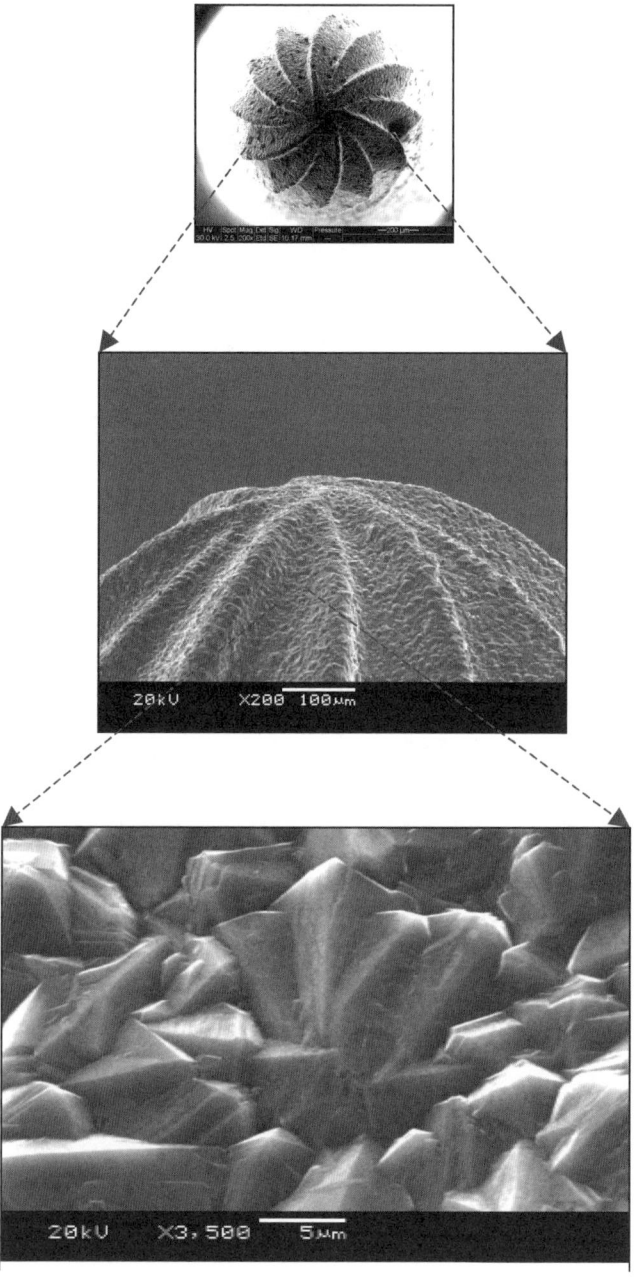

Figure 3.6. *(111) faceted octahedral shape diamond films on a microcutting tool*

Cutting materials for extreme requirements (for example, interrupted cuts or machining of high strength materials) consequently cannot be made from one single material, but may be realized by employing composite materials. Surface coatings may improve the tribological properties of cutting tools in an ideal way and therefore allow the application of tough or ductile substrate materials, respectively.

The coated microtools have been used to machine a number of materials including copper, aluminum, and iron alloys. The coated tools were compared with uncoated microtools to distinguish them in terms of their machining behavior. A micro-machining unit was specifically constructed at Purdue University for such a purpose with a maximum spindle speed of 500,000 revolutions per minute, feed rates of between 5 and 20 µm per revolution, and cutting speeds in the range of 100 to 200 meters per minute [79]. The micro-machining unit is shown in Figure 3.7. The machining center is constructed using three principal axes. each controlled using a DC motor connected to a Motionmaster™ controller. A laser light source is focused onto the rotating spindle in order to measure the speed of the cutting tool during machining. Post-machining analysis was performed using a scanning electron microscope to detect wear on the flanks of the cutting edges.

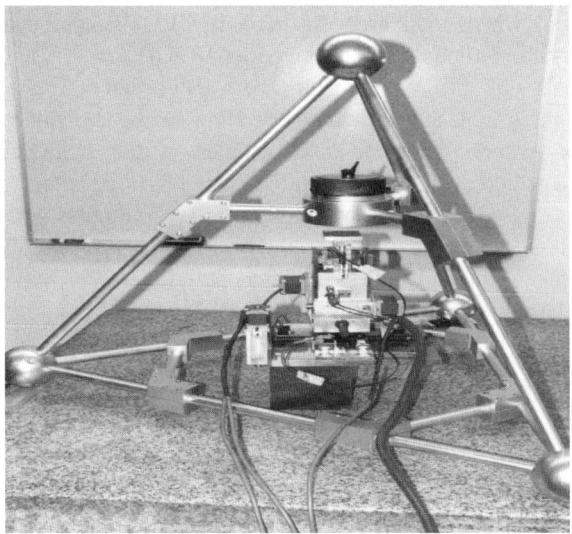

Figure 3.7. *Micromachining center using diamond microtools*

3.7.1. *Performance of diamond-coated microtool*

After machining an aluminum alloy material, very low roughness and chipping of the diamond-coated microtool were detected. A metal chip created from this machining operation. The chip clearly shows shear fronts separated by lamellae caused by plastic instabilities generated within the material at very high speeds. Diamond-coated and uncoated tools were compared by drilling a series of holes in the aluminum alloy. The wear of each tool was determined by examining the extent of flank wear. Uncoated tools appeared to chip at the flank face, and diamond-coated tools tended to lose individual diamonds at the flank face. Uncoated tools drilled an average of 8,000 holes before breakdown occurred, and the diamond-coated tools drilled an average of 24,000 holes [80]. In order to examine the cutting performance of the diamond-coated microtools, machining materials such as borosilicate glass, acrylic teeth and natural human teeth were used. A machining unit was set up for the laboratory microtool (AT23LR supplied by Metrodent, UK), which used to operate at 20,000-30,000 rpm with a feed rate of 0.2-0.5 mm/rev without water-cooling.

The flank wear of the microtools were estimated by SEM analysis at selected time intervals of 1 and 3 min. Prior to SEM analysis, diamond-coated microtools were ultrasonically washed with 6M sulfuric acid solution to remove any unwanted machining material, which could erode a surface of CVD diamond-coated microtool. For comparison, the commonly used conventional PCD (polycrystalline diamond) sintered microtools with different geometry were also tested on the same machining materials. These microtools are made by embedding synthetic diamond particles into a nickel matrix material to bond the particles at the cutting surfaces. The morphology of a sintered diamond microtool was observed after being tested on borosilicate glass at a cutting speed of 30,000 rpm for 5 minutes with an interval at every 30 sec. It is clearly evident that there is significant removal of diamond particles from the surface of the tool after 500 holes. As expected, there is deterioration in the abrasive performance of the PCD sintered diamond microtools.

After machining the diamond films are still intact on the pre-treated WC substrate and diamond coating displayed good adhesion. There is no indication of diffusive wears after the initial test for 500 holes. However, it was observed that the machining of materials such as glass pieces are eroded to the cutting edge of the diamond microtool as adhesive wear. After testing on acrylic teeth the wear mechanism probably involves adhesion as well as abrasion. Inorganic fillers from acrylic teeth, which adhered to the cutting tool surface in localized areas when a higher rate of abrasion, were used [81]. The uncoated WC-Co microtool displayed flank wear along the cutting edge of the microtool. The areas of flank wear were investigated at the cutting edge of the microtool. A series of machining experiments have been conducted using uncoated, diamond-coated microtools and sintered diamond microtools when machining extracted human tooth, acrylic tooth and borosilicate glass. The life of the microtools in the machining sense was compared by using the amount of flank wear exhibited by each type of microtool. The flank wear was measured at time intervals of 2, 3, 4, 5, 6 and 7 minutes of machining duration. Again, the microtools were examined using optical and scanning electron microscopic techniques and observed similar trends with microtools that were used in drilling experiments. It is evident that a longer duty cycle of machining could cause a higher rate of flank wear on the cutting edge of tool. Therefore, the cutting edge of the WC-Co microtool should have significant thickness of CVD diamond, which will enhance not only the quality of cutting but will prolong the tool life.

3.8. Conclusions

Thin film deposition technologies, particularly CVD and PVD, have become critical for the manufacture of a wide range of industrial and consumer products. Trends in historical developments in the CVD diamond suggest that the technology is highly likely to yield substantial benefits in emerging technological products for fields such as nanotechnology, biomedical engineering and MEMS devices. Several methods including plasma CVD, low pressure CVD and atmospheric pressure CVD have matured into processes that are routinely used in industry. Microwave and hot filament CVD methods are now commonly used to grow diamond and these can be modified to coat uniformly for tools, MEMS and biomedical applications. Diamond coatings examined on tools and biomedical tools showed significantly enhanced performance compared to uncoated tools.

3.9. References

1. Spear KE and Dismukes JP, in *Synthetic Diamond: Emerging CVD Science and Technology*, The Electrochemical Society, John Wiley & Sons Inc., New York, 1994.
2. Wentorf RH, "Synthesis of Diamond", *J. Phys. Chem.*, **69** (1965) 3063.
3. Butler JE and Woodin RL, "Characterisation and Analysis of Natural Diamonds", *Phil. Trans. R. Soc. Lond.*, **A342** (1993) 209.
4. Ashfold MNR, May PW, Rego CA and Everitt NM, "Raman Analysis of Diamond and Diamond Thin Films", *Chem. Soc. Rev.*, **23** (1994) 21.
5. Bachmann PK and Messier R, "Diamond Thin Films: Analysis and Applications", *Chem. Eng. News*, **67** (1989) 24.
6. Spear KE, "Chemical Vapour Deposition of Diamond", *J. Am. Ceram. Sci.*, **72** (1989) 171.
7. Joffreau PO, Haubner R and Lux B, "CVD Diamond and its Properties", *Mater. Res. Soc. Proc.*, **EA-15** (1988) 15.
8. Spitsyn BV, Bouilov LL and Deryagin BV, "Growth of Diamond by Chemical Vapour Deposition", *J. Cryst. Growth*, **52** (1981) 219.
9. Angus JC, "Chemical Properties of CVD Diamonds", *Electrochem. Soc. Proc.*, **89** (1989) 1

10. Yarbrough WA and Messier R, "Natural and Synthetic Diamonds", *Science*, **247** (1996) 688.
11. Messier R, Badzian AR, Badzian T, Spear KE, Bachmann PK and Roy R, "Growth of Thin Film Diamond by CVD Methods", *Thin Solid Films*, **153** (1987) 1.
12. Angus JC and Hayman CC, "Diamond Synthesis and Growth", *Science*, **241** (1988) 913.
13. Spear KE, "Chemical Vapour Deposition of Diamond", *Am. J. Ceram. Soc.*, **72** (1989) 171.
14. Kamo M, Sato U, Matsumoto S and Setaka N, "Growth and Formation of Diamond on Steel", *Cryst J. Growth*, **62** (1983) 642.
15. Saito Y, Matsuda S and Nagita S, "Growth of Diamond on Ceramic Substrates", *J. Mater. Sci. Lett.*, **5** (1986) 565.
16. Saito Y, Sato K, Tanaka H and Miyadera H, "Raman Spectroscopy of Diamond Films", *Mater. J. Sci.*, **24** (1989) 293.
17. Williams BE, Glass JT, Davis RF, Kobashi K and Horiuchi T, "Growth of Diamond in Vacuum", *Vac. J. Sci. Technol. A, Vac. Surf. Films*, **6** (1988) 1819.
18. Kobashi K, Nishimura K, Kawate Y and Horiuchi T, "Analysis of Diamond Films Grown Under Vacuum", *Vac. J. Sci. Technol. A, Vac. Surf. Films*, **6** (1988) 1816.
19. Liou Y, Inspector A, Weimer R and Messier R, "Studies of Diamond on Engineering Substrates", *Appl. Phys. Lett.*, **55** (1989) 631.
20. Zhu W, Randale CA, Badzian AR and Messier R, "Thin Film Diamond: Analysis and Characterization", *Vac. J. Sci. Technol. A, Vac. Surface Films*, **7** (1989) 2315.
21. Matsumoto S, "Effect of Growth Conditions on Diamond Films", *Mater. J. Sci. Lett.*, **4** (1985) 600.
22. Matsumoto S, Hino M and Kobayashi T, "Gas Concentration Effects in Diamond", *Appl. Phys. Lett.*, **51** (1987) 737
23. Vitkayage DJ, Rudder RA, Fountain GG and Markunas RJ, "Effect of Certain Gases on the Structure of Diamond", *Vac. J. Sci. Technol.*, **A6** (1988) 1812.
24. Meyer DE, Ianno NJ, Woolam JA, Swartzlander AB and Nelson AJ, "Current Research in the Growth of Diamonds", *Mater J. Res.*, **3** (1988) 1397.
25. Wood P, Wydeyen T and Tsuji O, in *Programs and Abstracts of the First International Conference on New Diamond Science and Technology*, New Diamond Forum, Tokyo, Japan, 1988.
26. Jackman RB, Beckman J and Foord JS, "Raman Spectroscopy of Diamond Thin Films and Coatings", *Appl. Phys. Lett.*, **66** (1995) 1018.
27. Suzuki K, Sawabe A, Yasuda H and Inuzuka T, "Characterization of Diamond by Positrons", *Appl. Phys. Lett.*, **50** (1987) 728.
28. Akatsuka F, Hirose Y and Kamaki K, "Effects of Hydrogen on Diamond Coatings", *Jap. J. Appl. Phys.*, **27** (1988) L1600.
29. Suzuki K, Sawabe A and Inuzuka T, "Effects of Gas Concentrations on Diamond Thin Films", *Jap. J. Appl. Phys.*, **29** (1990) 153.
30. Niu CM, Tsagaropoulos, Baglio J, Dwight K and Wold A, "Chemical Vapour Deposition of Thin Diamond Coatings", *J. Solid State Che*m., **91** (1991) 47.
31. Popovici G, Chao CH, Prelas MA, Charlson EJ and Meese JM, "Property Enhancements in Diamond", *Mater. J. Res.*, **10** (1995) 2011.
32. Chao CH, Popovici G, Charlson EJ, Charlson EM, Meese JM and Prelas MA, "Microcharacterization of Diamonds", *J. Cryst. Growth*, **140** (1994) 454.
33. Postek MT, Howard KS, Johnson AH and Macmichael KL, in *Scanning Electron Microscopy*, Plenum Press, New York, 1980.
34. Spirsyn BV, Bouilov LL and Deryagin BV, "CVD Diamond Films: Growth Studies", *J. Cryst. Growth*, **52** (1981) 219.

35. Kobashi K, Nishimura K, Kawate Y and Horiuchi T, "Growth of Diamond by Plasma Methods", *Phys. Rev. B*, **38** (1988) 4067.
36. Pickrell D, Zhu W, Badzian AR, Messier R and Newnham RE, "Effect of Substrate on the Growth of Diamond", *Mater. J. Res.*, **6** (1991) 1264.
37. Oatley CW, in *Scanning Electron Microscope*, Cambridge University Press, 1972.
38. Tobin MC, in *Laser Raman Spectroscopy*, Wiley Interscience, New York, 1971.
39. Colthup NB, Daley LH and Wiberley SE, in *Introduction to Infrared and Raman Spectroscopy*, Academic Press, New York, 1975.
40. Raman CV and Krishnan KS, "Spectroscopic Studies of Natural Materials", *Nature*, **121** (1928) 501.
41. Nemanich RJ, Glass JT, Lucovsky G and Shroder RE, "Selective Layer Deposition of Thin Films", *Vac. J. Sci. Tech.*, **6** (1988) 1783.
42. Knight DS and White WB, "Analysis of Diamond and Thin Films", *Mater. J. Res.*, **4** (1989) 385.
43. Solin SA and Ramdas AK, "Diamond Structure and Patterning During Slow Growth", *Phys. Rev. B*, **1** (1970) 1687.
44. Leyendecker T, Lemmer O, Jurgens A, Esser S and Ebberink J, "Coatings of Diamond on Ceramic Substrates", *Surf. Coat. Technol.*, **48** (1991) 253.
45. Murakawa M and Takeuchi S, "Diamond Synthesis on Tungsten Carbide, Properties of Bulk Crystalline Diamond", *Surf. Coat. Technol.*, **49** (1991) 359.
46. Yaskiki T, Nakamura T, Fujimori N and Nakai T, "Mechanisms of Ion-Induced Diamond Growth", *Surf. Coat. Technol.*, **52** (1992) 81.
47. Reineck J, Soderbery S, Eckholm P and Westergren K, "Selective Growth of Diamond Using Iron Catalysts", *Surf. Coat. Technol.*, **5** (1993) 47.
48. Wang HZ, Song RH and Tang SP, "On the Nature of CVD Diamond Growth", *Diamond and Relat. Mater.*, **2** (1993) 304.
49. Inspector A, Bauer CE and Oles EJ, "Diamond as a Cutting Tool Material", *Surf. Coat. Technol.*, **68/69** (1994) 359
50. Kanda K, Takehana S, Yoshida S, Watanabe R, Takano S, Ando H and Shimakura F, "Kinetic Analysis of Diamond Growth", *Surf. Coat. Technol.*, **73** (1995) 115
51. Luz B and Haubner R, in "Diamond and Diamond-like Films and Coatings", *NATO-ISI Series B, Physics*, **266**, edited by R. E. Clausing, 579, L.L. Horton, J.C. Angus, and P. Koidl, Plenum Press, NY, 1991.
52. Chen X and Narayan J, "Structure of (111) Diamond Films, Nucleation of Diamond on Graphite", *J. of App. Phys.*, **74**, (1993), 1468.
53. Klass W, Haubner R, and Lux B, "Interface Effects in the Growth of Diamond", *Diamond and Related Materials*, **6**, (1997), 240.
54. Zhu W, Yang PC, Glass JT, and Arezzo F, "Diamond Synthesis and its Effects", *Journal of Materials Research*, **10**, (1995), 1455.
55. Lux B and Haubner R, "Diamond and its Thermal Properties", *Ceramics International*, **22** (1996) 347.
56. *CRC Handbook of Chemistry and Physics*, edited by R.C. Weast, CRC Press, FL, 1989-1990.
57. Haubner R, Lindlbauer A, and Lux B, "Arc Discharge Growth of Diamond", *Diamond and Related Materials*, **2** (1993) 1505. 72
58. Chang CP, Flamm DL, Ibbotson DE, and Mucha JA, "Bias Enhanced Nucleation of Diamond", *J. Appl. Phys.*, **63** (1988) 1744.
59. Gusev MB, Babaey VG, Khvostov VV, Lopez-Ludena GM, Yu Brebadze A, Koyashin IY, and Alexanko AE, "Biased Growth of Diamond", *Diamond and Related Materials*, **6** (1997) 89-94.

60. Endler I, Barsch K, Leonhardt A, Scheibe HJ, Ziegele H, Fuchs I, and Raatz C, "Nucleation of Diamond Using Bias Enhanced Techniques", *Diamond and Related Materials*, **8** (1999) 834-839.
61. Kamiya S, Takahashi H, Polini R and Traversa E, "Performance of Diamond Coated Tungsten Carbide", *Diamond and Related Materials*, **9** (2000) 191-194.
62. Inspector A, Oles EJ and Bauer CE, "Diamond as a Cutting Materials", *Int. J. Refract; Metal Hard Materials*, **15** (1997) 49.
63. Itoh H, Osaki T, Iwahara H and Sakamoto H, "Novel Aspects of Diamond Growth", *J. Mat. Sci.*, **26** (1991) 370.
64. Liu H and Dandy DS, *Diamond Chemical Vapour Deposition*, Noyes, 1996.
65. Nazare MH and Neves AJ, *Properties, Growth and Application of Diamond*, INSPEC, 2001.
66. Zhang GF and Buck V, "Surface Coated Diamond Substrates", *Surf. Coat. Technol.*, **132** (2000) 256.
67. Haubner R, Kubelka S, Lux B, Griesser M and Grasserbauer M, "TiN Coated Diamond Thin Films", *J. Physics. 4th Coll*, **C5**, 5 (1995) 753.
68. May P, Rego C, Thomas R, Ashfold MN and Rosser KN, "Nucleation and Growth of Interlayers on Diamond Films", *Diamond and Related Materials*, **3** (1994) 810.
69. Gouzman I and Hoffmann A, "Performance of Coated Diamond Films", *Diamond and Related Materials*, **7** (1998) 209.
70. Wang W, Liao K, Wang J, Fang L, Ding P, Esteve J, Polo MC and Sanchez G, "Characterisation of Diamond Thin Films Coated with Titanium Nitride", *Diamond and Related Materials*, **8** (1999) 123
71. Wang BB, Wang W and Liao K, "Effect of Thin Coatings Attached to Diamond", *Diamond and Related Materials*, **10** (2001) 1622.
72. Kim YK, Han YS and Lee JY., "Diamond Tools Coated with Sensitive Materials", *Diamond and Related Materials*, **7** (1998) 96.
73. Wang WL, Liao KJ and Gao GC, "Diamond Formation in Vacuum and Air", *Surf. Coat. Technol.*, **126** (2000) 195.
74. Polo MC, Wang W, Sanshez G, Andujar J and Esteve J, "Spectroscopy of Diamond Thin Films Bonded to Ceramic Substrates", *Diamond and Related Materials*, **6** (1997) 579.
75. Kamiya S, Yoshida N, Tamura Y, Saka M and Abe H, "CVD Diamond Coatings and the Applications", *Surf. Coat. Technol.*, **142-144** (2001) 738.
76. Sein H, Ahmed W, Rego CA, Jones AN, Amar M, Jackson M J, Polini R, "Chemical Vapour Deposition Diamond Coating on Tungsten Carbide Dental Cutting Tools", *J. Phys: Condens. Matter*, **15** (2003) S2961.
77. May PW, Rego CA, Thomas RM, Ashford MNR and Rosser KN, "Nucleation and Growth of Interlayers on Diamond Films", *Diamond and Related Materials*, **3** (1994) 810.
78. Amirhaghi S, Reehal HS, Plappert E, Bajic Z, Wood RJK, Wheeler DW, "TiC and TiN Coated Diamond Substrates", *Diamond and Related Materials*, **8** (1999) 845.
79. Jackson M J, Gill MDH, Ahmed W, Sein H, "Manufacture of Diamond Coated Cutting Tools for Micromachining Applications", *Proceedings of the Institute of Mechanical Engineers – (Part L): J. Materials*, **217** (2003) 77.
80. Sein H, Jackson M J, Ahmed W, Rego CA, "Deposition of Diamond Coatings to Micro Machining Cutting Tools", *New Diamond and Frontier Carbon Technology*, **12**(6) (2000) 1.
81. Sein H, Ahmed W, Jackson M J, Woodwards R and Polini R, "Performance and Characterisation of CVD Diamond Tools for Dental and Micromachining Applications", *Thin Solid Films*, **447-448** (2004) 455.

Chapter 4

Conventional Processes: Microturning, Microdrilling and Micromilling

4.1. Introduction

4.1.1. *Definitions and technological possibilities*

Micromachining is the most basic technology for the production of miniaturized parts and components [BYR 03]. It includes *bulk* micromachining processes which produce structures inside a substrate and *surface* micromachining processes which are based on the deposition and etching of different structural layers on top of the substrate. On the other hand, in "mechanical/conventional" micromachining the material removal process resembles macroscopic machining processes such as drilling, milling and others.

From such a point of view, micromachining encompasses micro-electro-mechanical systems (MEMS), microsystems technologies (MST) and, in addition, includes processes related to the production and packaging of microsystems [MAS 00a]. Micromachining by precision technology such as 3D micro-EDM, micro-laser machining, microcutting, microgrinding, etc. can produce microscopic and mesoscopic mechanical structures of complex shapes.

Chapter written by Wit GRZESIK.

The word *micromachining* may literally mean the machining of the dimensions between 1 μm to 999 μm but it should also consider the contemporary levels of conventional technologies. For instance, the micro-macro border was set around 200 μm with some variation among different manufacturing methods [BYR 03]. On the other hand, if micro likewise means "very small", it should also indicate too small to be machined easily by normal machining. In such a sense the range of micromachining can be placed between 1 μm and 500 μm [BYR 03].

Although micromachining and conventional precision machining use many of the same techniques and equipment, the goals are quite different. Precision engineering is informally defined as the production of parts with a size-to-tolerance ratio greater than 10,000:1. In micromachining, the goal is to produce mini- and microsystems (typically 1 or less cm in size) with microcomponents (typically 1 to 100 μm in size) with sufficient tolerances to achieve required functionality and repeatability [DOR 94].

4.1.2. *Main applications of micromachining*

The range of machining accuracy between 0.01 μm (lower range of precision manufacturing) and 0.001 μm (1 nm) is termed by Taniguchi as *ultra-precision machining* [BYR 03]. Figure 4.1 shows the capability of conventional micromachining in relation to other manufacturing processes such as laser machining, EDM, grinding and the LIGA (photo-lithography method using a synchrotron) process. It is worth noticing that micromachining makes it possible to attain Ra values in the range down to almost 5 nm for features down to 1 μm. In contrast, as specified by Taniguchi, for "normal machining", e.g. CNC turning and milling machines, accuracies of 10 μm to 100 μm can be achieved.

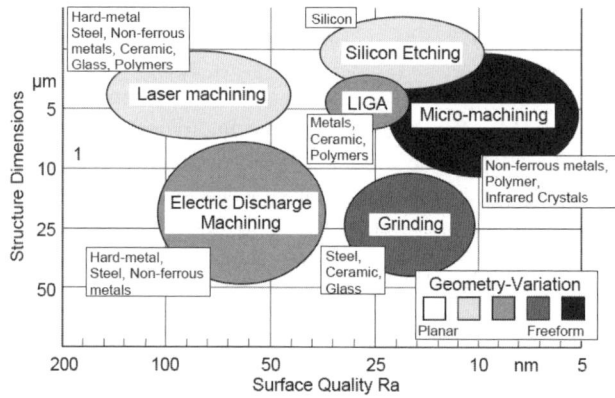

Figure 4.1. *Micromachining relative to other machining processes [BYR 03]*

Currently, the following areas of interest can be observed [ARON 04, ARON 07, BIB 06, CHA 06]:

1) The medical and electronic industries are the most active in this area and in both cases the incentives are the same: making smaller parts with more capabilities. The cell phone is an obvious example and the goal is to cram more features into one unit.

2) The medical industry is under pressure to produce various repair and pain-relief devices which must be small enough to embed into the human body. Micro-sized components can lead to the progression of minimally invasive techniques for orthopedic, cardiology and neurology applications.

3) The aerospace industry is a large end-user for fasteners, fittings, sensors and various flow-control device.

4) The automotive sector introduces many innovations for convenience, entertainment and safety items every year. As a rule, many of these features require very small motors and actuators. From a construction point of view, there is a trend to install complex fuel injection and small flow-control elements. Some of the most representative examples in this category are micro-valves and fluidic graphite channels (circuits) for fuel cell applications.

Companies now involved in micromanufacturing originally came from three sources [ARON 07]. First, a number of companies have been successful partners in the micromarket for some time, such as those involved with microturning using Swiss lathes. The second group consists of manufacturers who have modified their equipment to meet the special needs of micromanufacturing. For example, many tool builders have generated lines of small tools fairly recently. The third is created by entrepreneurs who, in the last few years, introduced totally new products for the micromarket such as desktop machine tools and robots.

In the following sections some representative technologies and examples of parts shaped by means of various micromachining techniques are selected in order to image the scale of their miniaturization from millimeters to micrometers and associated manufacturing limitations.

74   Nano and Micromachining

4.2. Microturning

4.2.1. *Characteristic features and applications*

As mentioned in section 4.1.2 various Swiss-type machine tools, *ultra-precision* lathes for diamond turning and miniature desktop machines, are used to perform microturning operations on parts made of different materials.

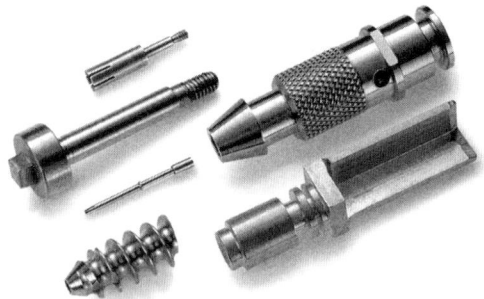

Figure 4.2. *Typical parts machined by Swiss-type machine from Figure.4.7 [CIT 07]*

According to today's experiences, high-precision Swiss-type turning centers can produce a variety of tiny precision parts with a maximum machining diameter of 7 mm (5/32") and length of 40 mm (double diameter per chucking). Some examples of such small diameter parts with special geometric features (nozzles, shafts, EDM copper electrodes, single-thread injector worm screws) produced by combination of turning, drilling, milling and threading operations are presented in Figure 4.2.

Figure 4.3. *Diamond turning of optical quality face surface [RAH 07]*

The next application group of microturning operations with *ultra-precision* mode makes it possible to machine materials from a few microns to sub-micron level to achieve ductile mode machining on hard-to-machine materials such as electroless nickel plating, silicon, quartz, glass and ceramics without subsurface defects. Such a machining process is easily able to produce mirror surfaces of less than 10 nm surface finish and form error of less than 1 μm on some diamond turnable materials [RAH 07]. As shown in Figure 4.3, one of the major applications of *ultra-precision* (diamond) turning is machining of electroless nickel plated molding dies for plastic optical parts such as LCD projection TVs. As reported in [RAH 07] surface roughness produced on these diamond-turned mirror surfaces was achieved to be less than 6 nm in Ra.

4.2.2. Microturning tools and tooling systems

Currently, the leading cutting tool manufacturers offer a wide range of miniature tools and appropriate clamping systems. They are used as a tool block of (placed in appropriate order as a gang of tools) systems dedicated to Swiss-type lathes or multi-purpose cutting systems mainly for simultaneous boring and grooving/part-off operations. Larger versions of the sliding head machines are also equipped with turrets as well as a gang tool to give further flexibility for using standard tool holders as well as quick change tooling [SAN 07]. For instance, Figure 4.4 shows very small (trade mark Supermini® by Horn, USA) coated monolithic carbide tools to machine precise small bores with the smallest diameter of 0.2 mm (0.008"). In addition, CBN (from diameter of 3 mm) and PCD (from diameter of 4 mm) tipped inserts are also available for boring operations in cast iron or hardened materials [HOR 08].

a) b)

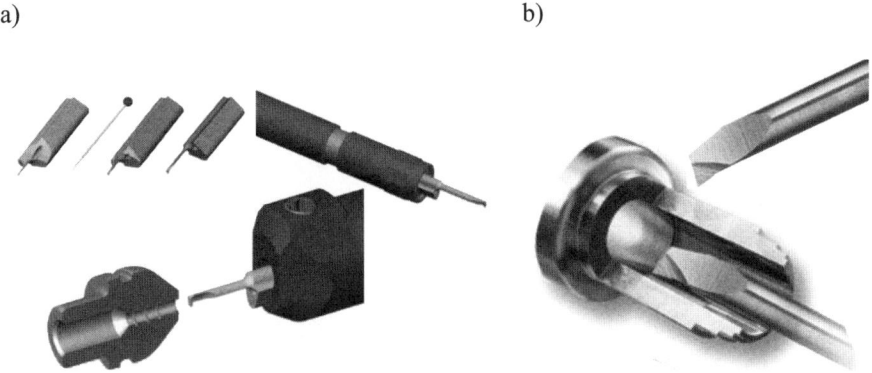

Figure 4.4. *Miniature tools for grooving and boring on lathes:*
a) Supermini® [HOR 08] and b) CoroTurn XS [SAN 07]

Sandvik Coromant [SAN 07] introduced CoroTurn XS program of boring inserts and holders (see Figure 4.4b) designed for turning, copying, grooving, pre-parting, profiling and threading of components with bores down to 1 mm. In addition, a number of external tools are capable of performing the appropriate operations on extremely small parts. The Kennametal's concept of a special tooling system named KM Micro [KEN 07] is shown in Figure 4.5. It enables machining operations such as turning, boring, cut-off and drilling. The system consists of a clamping unit and a cutting unit.

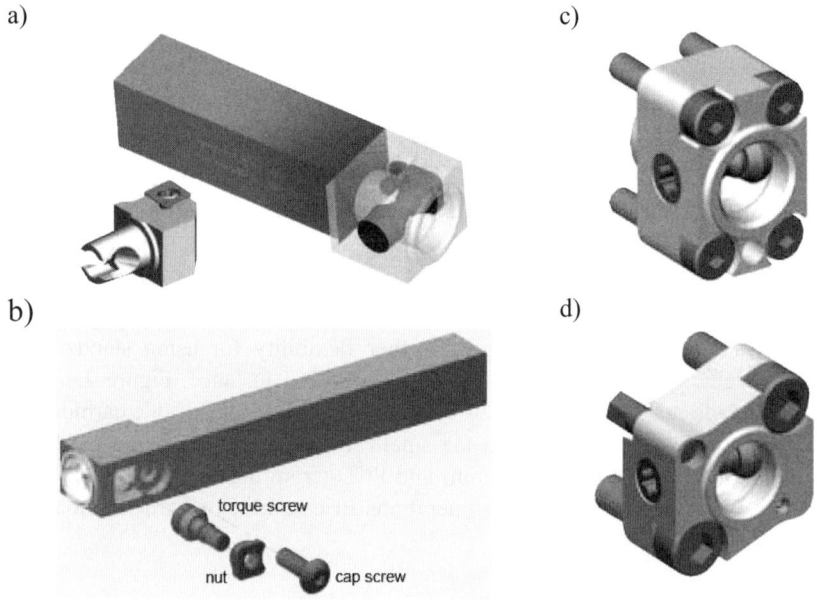

Figure 4.5. *KM Micro Quick-Change Tooling [KEN 07]:
a) clamping system and micro quick-change cutting unit, b) square shank adapter,
c) universal flange clamping unit, d) flange attachment unit*

The quick-change cutting unit can be simply replaced by a cutting unit that contains a new insert. The clamping unit can be a square shank (Figure 4.5b) or a round shank. Bolt-on flanges (Figure 4.5c and d) are also available to be used with a special mounting block. Their use on Swiss-type machines often makes it possible to add one or more additional positions in the existing tooling envelope and, as a result, machine uptime decreases at a rate of 25-40%. It should be noted that both square and round shanks can be used in any tool slot, turret or block with a dimension matching the system size.

In turning of small-diameter parts with high aspect ratio, a critical issue is the elastic deflection of the slender workpiece. When using the automatic Swiss-type lathes/centers, a special material loading device shown in Figure 4.6, in which the tools clamped in the head (6) circulate instead of the workpiece, can reduce the elastic deflection. The tools can also move in the radial direction with a feed fr. According to the scheme from Figure 4.6, the raw material is a wire-like bar precisely drawn for reducing the dimensional tolerance necessary for minimizing a radial displacement between the material and the supporting sleeve (4). During machining the wire is moved and rotated as a coil by a collet and precisely grasped and fed into the machine. A close support of the material against the tool contact point reduces the length of the unbounded force arm ld. As a result, parts of the smallest diameter of 0.3 mm and the length of 15 mm (the aspect ratio equal to 50) can be precisely machined.

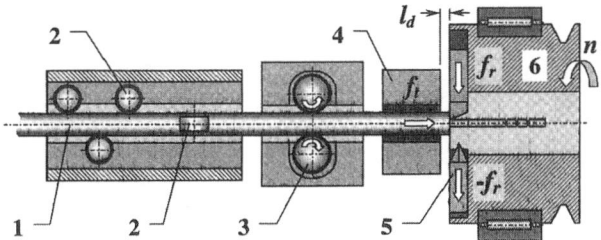

Figure 4.6. *Material feeding device with rotating tool head: 1) raw material, 2) straightening unit, 3) longitudinal feed drive, 4) supporting sleeve, 5) cutting tool, 6) rotating tool head*

4.2.3. Machine tools for microturning

In general small parts can be produced using CNC-lathes, sliding head machines and older automatics. The early manufacture of clocks and watches in Switzerland stimulated the development of machine tools (called Swiss-type lathes) capable of making a variety of small parts with the necessary precision. Later, these machines evolved into Swiss-type turning centers with milling, drilling and boring capabilities. The newer trend is to design machines with the ability to finish small parts (currently 4 mm in diameter) in a single setup with lower cost. The most up-to-date method of machining small components today is by using sliding head-stock machines because they are the most efficient way of producing large volumes. This name is originated from the fact that the sliding head-stock machine is equipped with a main head function, which use s a guide bush in order that the material can slide through in the Z-direction, leaving the tools stationary so that high stability is achieved. Most sliding head machines are CNC-controlled and take a maximum of 32 mm bar diameter [SAN 07]. They are nearly always fitted with a bar-feeder to maximize the productivity of the automated process. Figure 4.7 shows a high-speed, high-precision lathe for machining of tiny precision parts with a maximum machining diameter of 7 mm and length of 40 mm (Figure 4.2 in section 4.2.1), but a new model can go down to subminiature parts with 0.5 mm in diameter. The machine can perform metal cutting at speeds of up to 12,000 rpm but without the rotary guide bushing the cutting speeds go up to 16,000 rpm. Sub-spindles for drilling and milling deliver 10,000 rpm. The use of linear motors to drive the slide and tool post results in fast part processing and eliminates thermal distortion. A scale feedback control system for all axes offers the perfect machining configuration for small, high precision parts. The reduction of non-cutting idle time was achieved by the possibility of opening and closing the chuck without deceleration of the spindle.

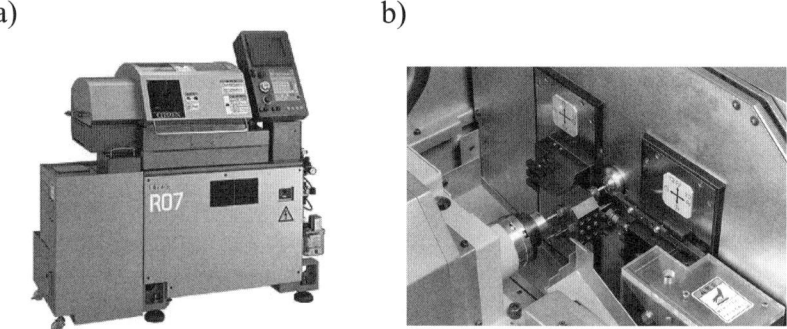

Figure 4.7. *(a) CNC automatic micro lathe and (b) axis arrangement [MAR 07]*

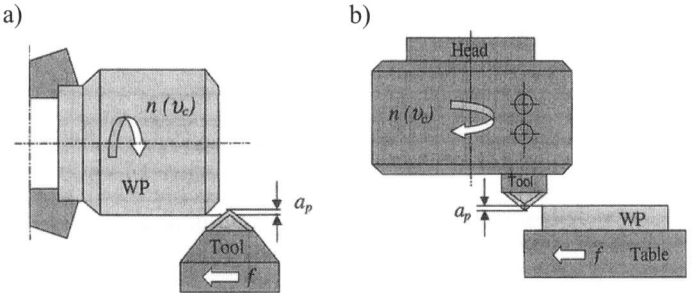

Figure 4.8. *General kinematic schemas of (a) diamond and (b) fly-cutting turning*

This effect was increased by installing two independent tool posts (Figure 4.7b), which eliminate tool exchange. The machine can be optionally equipped with a vacuum part removal system that delivers parts segregated from the chips and carousel workpiece separator, which greatly improves productivity. Rotary tools mounted into the gang tool post allow small-diameter parts requiring complex machining such as polygon turning and end face drilling, to be machined easily. Citizen sliding head machines are equipped with a Quick Start (QS™) holding system, consisting of a number of holders in the tool post with special spring loaded clamping, developed by Sandvik for external machining [SAN 07].

As shown in Figure 4.8, mirror-like surfaces can be turned using ultra-precision lathes with (a) conventional kinematics and (b) machines with fly-cutting tools in which the kinematics is similar to conventional vertical milling. Rigorously, the machining operations are performed on the very rigid machine tools equipped with ultra-precision linear drives and solid polymer concrete bases, which eliminates vibration and ensures the industry leading accuracy of 1 µm (0.0001"). This solid machine bed provides safe, reliable processing, even at very high cutting speeds of 15,000 rpm [MAI 2008].

4.3. Microdrilling

4.3.1. *Characteristic features and applications*

Microdrilling is capable of producing holes several tens of micrometers in size for many large-scale practical applications such as printed circuit boards, fuel nozzles and spinners for the textile industry. A particular example of precise microdrilling demands is the production of the head for ink-jet printers in which a single row of 251 holes of 50 µm diameter has to have the exact pitch of 200 µm. In general, holemaking operations with miniature drills, popularly termed microdrills, are performed for hole diameters ranging below 0.5 mm. In contrast, the minimum diameter of commercially available microdrills is in the range of 5-50 µm depending on the basic shape and tool material used.

The advantages of microdrilling are as follows [MAS 00b]:

(1) Most metal and plastics, as well their composites can be machined easily because the electrical properties of the workpiece do not influence the process (vide EDM).

(2) Machining time can be controlled effectively due to satisfactory stability of the process.

On the other hand, some disadvantages occur, namely:

(1) Produced holes are often inclined and in some cases correct positioning is necessary to avoid excessive straightness errors.

(2) The brittle chipping of the workpiece and the breakage of microdrills limit the application of the conventional holemaking in very hard and brittle materials. An alternative is the use of diamond drills.

4.3.2. Microdrills and tooling systems

As reported earlier, several of the major cutting tool manufacturers offer tool systems specifically dedicated to the performing micromanufacturing on Swiss-type or small CNC machines [GRZ 08]. The bit material ranges from HSS (more resistant to breaking) to carbide (stiffer but more brittle). Iscar offers tools with multidirectional grooving inserts as narrow as 0.5 mm (0.0197"), a multi-functional carbide tool of 4 mm in diameter (named by PICCO-MFT) for drilling, facing, internal and external turning and threading, and miniature carbide coolant-fed drills and ISO turning tools with shanks as small as 10 mm (0.375"). Mitsubishi Carbide produced the smallest solid micrograin carbide, with coolant-through drills 1-2.8 mm (0.040-0.110") in diameter for drilling in aircraft aluminum, 6Al4V titanium, 304L stainless steel and 4130 steel. The drills are coated with a PVD-Al,TiN layer with high resistance to adhesive welding. Moreover, MA Ford offers off-the-shelf drills down to 0.1 (0.05) mm (0.0039"(0.002")) diameter and die-mold ball endmills of 0.4 mm (0.0156") and traditional four-flute endmills of 0.12 mm (0.0050") in diameter. Some types of microtools dedicated to HSM, including solid micrograin carbide drills, are shown integrally in Figure 4.9. Moreover, their characteristic dimensions and typical applications are listed in Table 4.1.

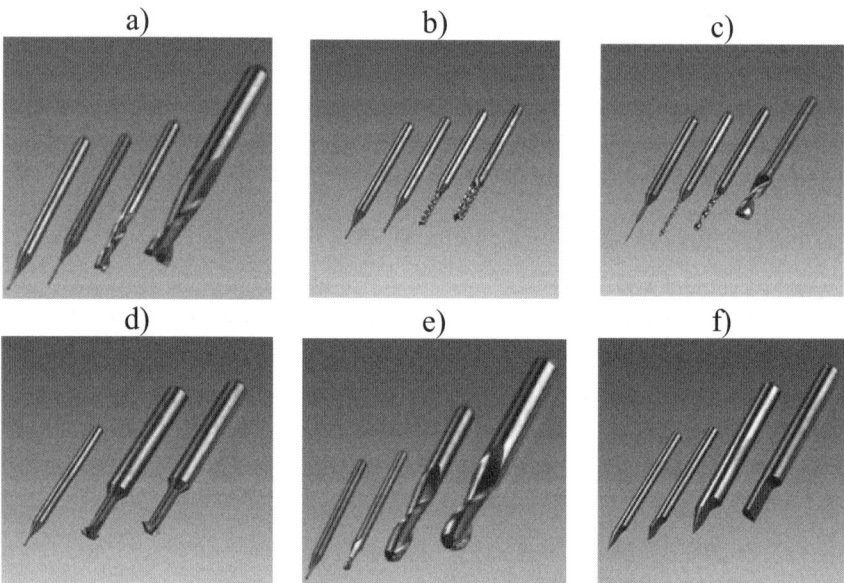

Figure 4.9. *Microtools for HSM [DAT 07]: a) double flute endmills, b) fish tail endmills for PCB milling, c) micrograin carbide drills with 1/8" shank, d) thread mills, e) ball nose endmills, f) standard engraving tools*

Type of microtool	Diameter in mm	General use
Double-fluted endmills	0.3-1.5	Aluminum, brass, duro-plastics, copper
Fish tail endmills for PCB milling	0.6-3.0	Printed-circuit material, e.g. FR4
Micrograin carbide drills	0.4-3.0	Aluminum, brass, plastics, copper, etc.
Thread mills	3 and 6	Non-ferrous materials
Ball nose endmills	3, 6 and 8	3D and modelmaking applications
Standard engraving tools with conical tool-tip	3 and 6	Aluminum and plastics

Table 4.1. *Characteristics of microtooling from Figure 4.9 [DAT 07]*

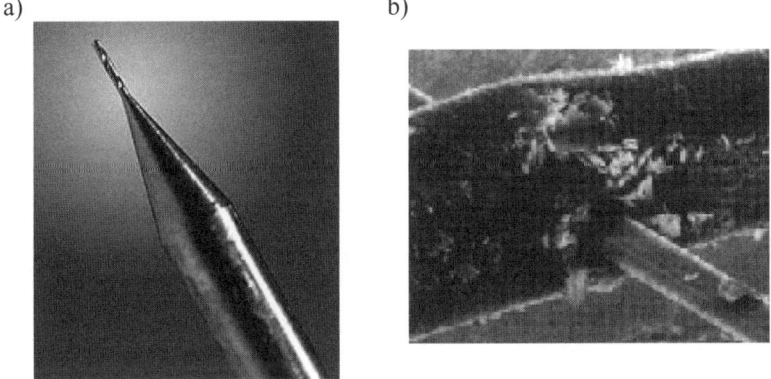

Figure 4.10. *HSS spiral microdrill (a) in comparison with a human hair (b) [MIN 08]*

Figure 4.10 illustrates the 0.03 mm (0.0012") spiral HSS microdrill which was especially designed for the smallest diameter microdrilling with excellent chip clearing properties against a human hair. Figure 4.10b depicts a human hair, approximately 0.003" in diameter, with a hole drilled in it by a 0.0012" diameter drill. It should be noted that drills in this micro-range require the ultimate in microdrilling equipment that assures minimum drill runout. For this reason a special microdrilling machine (see Figure 4.13), where precision Microdrills are held in special V-shaped diamond bearings virtually eliminating drill runout, has been developed.

82   Nano and Micromachining

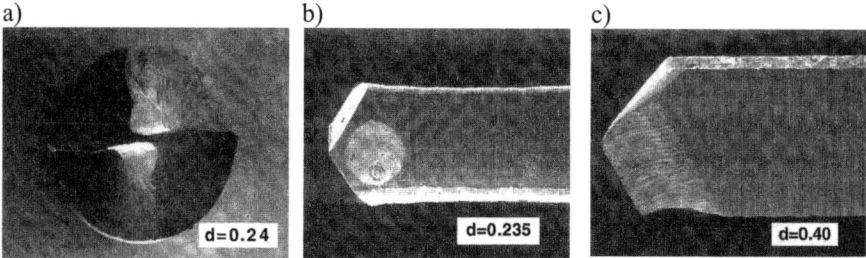

Figure 4.11. *Microdrills with modified geometry: a) twist drill with undercut chisel edge, b) spade drill with shell-like cavity on rake surface, c) D-shape drill with groove parallel to major cutting edge [KUD 07]*

The majority of microdrilling operations are performed using twist drills, although the smallest are spade and D-shape microdrills, because of their better performance. As a result, a spade, rather than a spiral tip, is recommended in the range from 0.001 to 0.005" (0.025-0.13 mm) [MINI 08]. The basic shape of microdrills can be additionally modified by analogy to conventional monolithic drills of large diameters as shown in Figure 4.11. As proven in practice, such modifications cause the cutting forces to diminish and the hole quality improves.

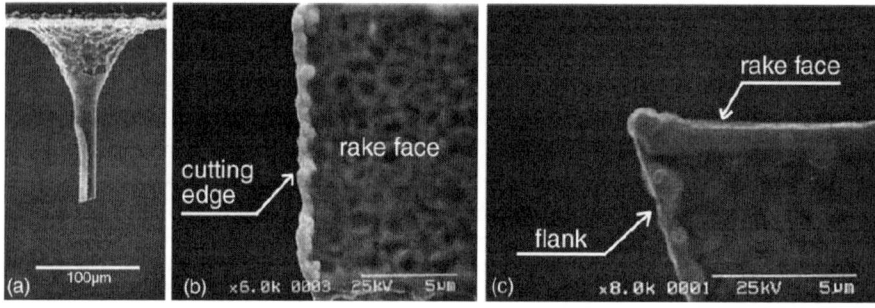

Figure 4.12. *Microdrill of 17µm diameter a fabricated by electro-discharge grinding (WEDG): a) general SEM view, b) close-up view to the rake face, c) close-up view parallel to the cutting edge [EGA 02a]*

Microdrills of the smallest obtainable diameter of approximately 6 µm with a D-shape cross-section and a cutting edge radius of 0.5 µm can be fabricated by wire electro-discharge grinding (WEDG) as reported in [EGA 02a]. In order to realize the ductile mode cutting of hard and brittle materials like monocrystalline silicon, the cut depth should be kept at 0.1 µm. In the case of fabricated microdrill of 17 µm diameter, shown in Figure 4.12, it is necessary for the clearance angle to be larger than 0° to avoid fractures at the hole entrance. It was documented that the feed rate of 0.3 µm/s is possible without tool damage when using a 20 µm diameter and 50 µm length drill.

4.3.3. Machine tools for microdrilling

The most popular manufacturing technologies of microhole making are mechanical drilling and lasers and micro-EDM. Conventional microdrilling is performed using multi-purpose CNC controlled machining center and a Citizen Swiss style double spindle lathe (these machines employ high-speed spindles for the efficient production of very small parts), as well as conventional or specially designed micro-drilling machines. The challenge is to drill holes down to 0.0005" accurately and consistently through a wide range of materials, including most metals and plastics. Both round or shaped microscopic holes can be produced keeping tolerances within 0.0001". Precise diameter holes (± 0.0001") are drilled in nozzles, orifice plates, dies, spinnerettes and many other parts for several different industries. It is important to produce burr-free holes in a wide variety of materials.

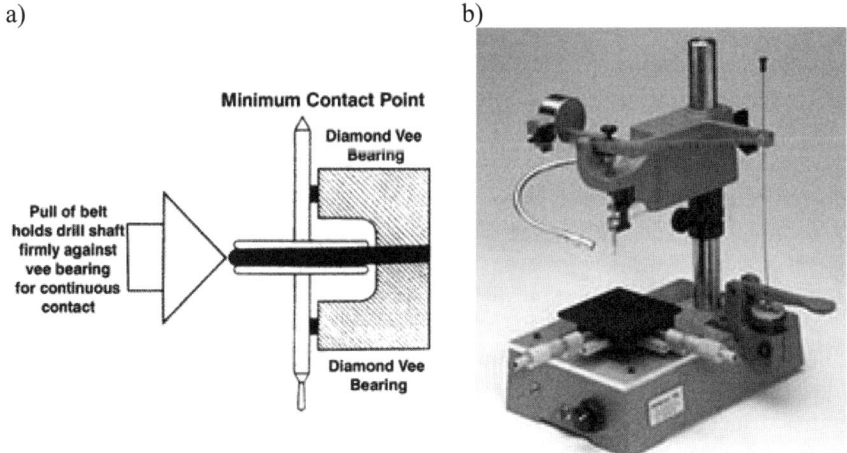

Figure 4.13. *Microdrilling machine: (a) a concept (b) and real view [MIN 08]*

Figure 4.13 shows microdrilling machines which represent a completely new and innovative approach to precision micro-drilling. Super hard in-line interchangeable V-shaped diamond bearings (Figure 4.13a) replace the conventional ball or sleeve bearings. The drills are mounted within an interchangeable drive pulley and held against the vee bearing by means of a special endless drive belt. The drill now becomes its own spindle, eliminating the runout that is a common factor in conventional machines using ball bearings, sleeve bearings, spindles, collets or chucks. Manual microdrilling machines (Figure 4.13b), for precision drilling of microscopically small holes in sizes from 0.025 to 0.50 mm diameter (0.001" to 0.020"), as well as CNC-controlled (automatic) microdrilling systems, are produced. Speeds around 3,500 rpm are recommended, with 5,000 rpm maximum. Drills, endmills and special tooling are also manufactured using a proprietary tool grinding technique.

Figure 4.14. *Special air bearing spindle for microdrilling machines [ARO 04]*

Sometimes customers want a precision that will allow them to eliminate extra steps or hand finishing. One of the goals, chiefly among researchers, is the 500,000 rpm air bearing spindle, for high-speed holemaking shown in Figure 4.14. Currently, the maximum speed for an air-bearing spindle is 180,000 rpm, and for a conventional ball bearing spindle it is 160,000 rpm. It is a 60 mm spindle with a 1/8" (3 mm) collet capacity. However, we are pushing the envelope to increase speed.

In the design of ball bearing spindles, power is rapidly increasing due to a couple of developments. First, permanent magnet motor design is improving. They offer a significant increase in power compared to typical ac-induction motors. Designers have found ways to keep the magnets on the shaft at higher speeds. The other improvement is thermal stability in the shaft. There is a much smaller temperature difference in the shaft between idle and full-load operating conditions. The reduced temperature fluctuation means a more stable tool.

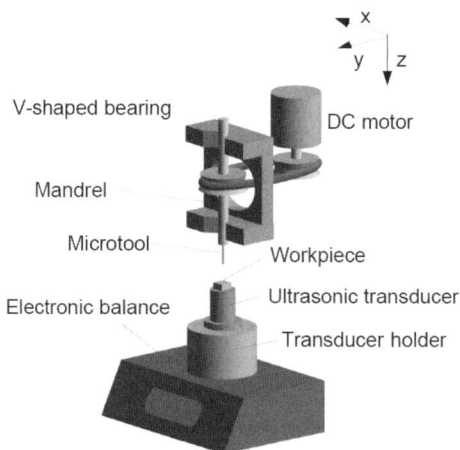

Figure 4.15. *A setup for ultrasonic vibrations microdrilling [EGA 02b]*

Section 4.3.2 mentions that the ductile mode cutting of brittle materials such as polycrystalline silicon or glass is possible when using the cut depth usually less than 1 μm. A large cut depth or high penetration increases the cutting force, leading to tool breakage or transition from the ductile mode to the brittle mode. In such cases the assistance of ultrasonic vibration is a useful technique for the reduction of cutting force [EGA 02b]. When using the microdrill shown in Figure 4.15 with an ultrasonic transducer assembled on the workpiece, which generates vibration of the amplitude 0.8 μm, it was possible to produce microholes of 10 μm in diameter in glass and obtain the ductile mode at a cut depth of 0.05 μm. It should be noted that the microtool was fabricated by the same method as that shown in Figure 4.12. As a result, due to 60-70% reduction of the cutting force, microholes were without fractures or cracks and the permissible penetration was increased to 0.05 μm/s.

4.4. Micromilling

4.4.1. *Characteristic features and applications*

Mechanical micromilling is not generally applied for micromachining due to substantial difficulties in obtaining appropriate microtools [MAS 97]. A comparison between capabilities of the micromilling process and other non-mechanical micromachining processes is given in Table 4.2.

	Micromachining capabilities				
	High Speed Milling	Sinker EDM	Wire EDM	X-ray Lithography	Ion Beam Machining
Minimum structure size (μm)	50	5 to 10	15-20	-	< 0.1
Surface finish Ra (μm)	1	0.2	0.05	-	0.04 to 0.15
Inner radius (μm)	50	< 10	~ 15	-	0.01
Aspect ratio	100-150	~ 20	100-150	100	10
Drawback	Heat fracture	Slow removal rates	Through shapes only	Learning curve to moldmakers	Learning curve to moldmakers

Table 4.2. *Listing of micromachining techniques and their capabilities in comparison to high-speed milling (HSM)*

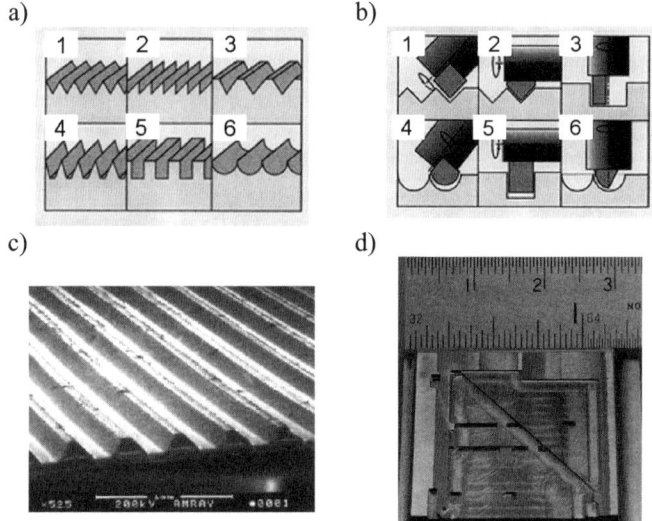

Figure 4.16. *Milling of ultra-precision microgrooves with single crystal diamond endmills [GRZ 08] and milled complex microstructure [ARO 04]*

However, precision micromilling has many various realizations depending on the tool shape and the kinematic features. The most complex is 3D micromilling of cavities in miniaturized moulds (Figure 4.16d), chemical microreactors and fluidic parts. Fly milling is basically a method for producing grooves when they are cut in different directions. It is particularly effective for producing a large number of microcolumn arrays.

Single crystal diamond with the cutting edge radius of about 50 nm is a preferred tool material for conventional microcutting (a 100 μm size features are routinely machined) of non-ferrous materials like brass, aluminum, copper and electroless nickel. Figure 4.16a shows microgrooves of different shapes (numbered as 1-7) produced by means of an *ultra-precision* milling machine and single crystal diamond mills with different operating arrangements (Figure 4.16b). This method allows us to produce V-shaped grooves with high shape accuracy without burrs and surface roughness of 48 nm Rz (P-V) value. Another application example for micro-diamond machining are flow passages for micro-compact heat exchangers shown in Figure 4.16c, in which the surface enhancement performed is necessary to reduce convection boundary resistance. Each of all passages from Figure 4.16c is 100 μm in width at the bottom and 80 μm in depth. The material is high conductivity copper foil of 125 μm thick. Micro-diamond machining can also be used to complement lithography process. As reported in [WUE 01] tungsten carbide tools with extremely small cutting edge radii can produce reproducible surface roughness lower than Rz=0.8 μm when micromilling of miniaturized steel components. Recently, special methods of ultra-precision cutting of a carbon steel parts with diamond cutting tools using cryogenic cooling and ultrasonic vibrations have also been proposed [BRI 01].

4.4.2. *Micromills and tooling systems*

The size of precision micro-cutting tools determines the size limit and accuracy of microstructured features. Diamond tools are often used for ultraprecision machining (see Figure 4.16) but they are limited to machine non-ferrous materials. Therefore, microtools such as micro-endmills and drills are generally made from tungsten carbides with grain size smaller than 600 nm [FLE 04].

Micromills can have a number of different, but simple cross-section profiles as shown in Figure 4.17. The first two profiles are used in D-shape and spade microdrills in which the cutting part is much shorter (lcp≈2d). In addition, they have a flat front plane and the clearance angle on the major cutting edges equals 0. Next, two cross-sections have regular polygonal forms and a large negative rake angle, between -45° and -60°. The side surfaces in such microdrills can be shaped as the concave surfaces which improve the cutting performance. It is reported by [KUD 07] that slotting micromills with the minimum width of about 50 μm are similar to micro-endmills, which are widely used in the micromachining industry. Some of them of a diameter of 2 mm and the width of 30 μm are uniquely fabricated from diamond-like carbon (DLC) layers using the CVD technique.

88   Nano and Micromachining

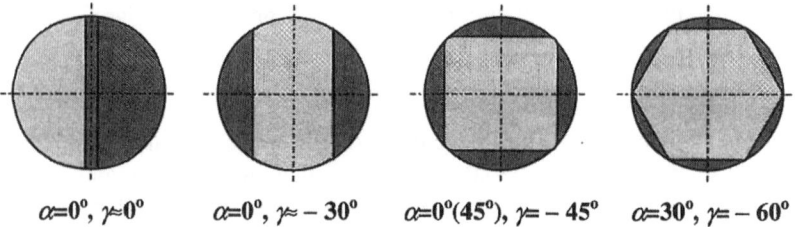

$\alpha=0°, \gamma=0°$ $\alpha=0°, \gamma=-30°$ $\alpha=0°(45°), \gamma=-45°$ $\alpha=30°, \gamma=-60°$

Figure 4.17. *Cross-section profiles of smallest end micromill [KUD 07]*

The tiny milling, shown in Figure 4.18, capable of generating features that are at least 25 μm in size, are made by sculpting carbide and HSS blanks using a focused ion beam. Endmills made in this way can be as small as 20 μm in diameter and turning tools can be about 10 μm in width. These tools were used in the meso-machining of miniature nuclear weapon components made of aluminum, brass and 4340 steel [ZEL 07].

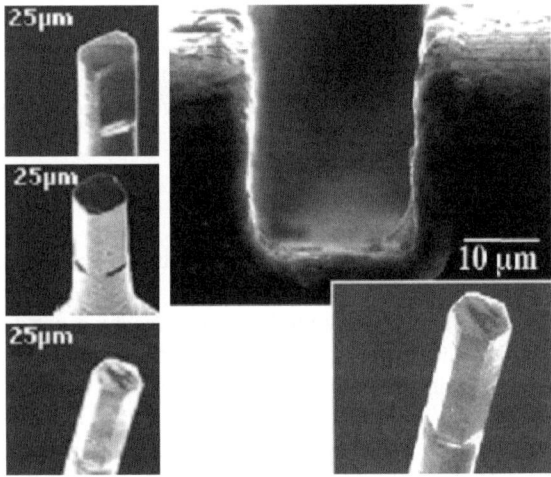

Figure 4.18. *A range of micro-endmills (three left photos) and magnification of a 25 μm endmill (right photo) [GRZ 08]*

When shaping tools with an ion beam, there was no way to generate the fluted, complex geometry which is typical of a standard endmill. Instead, meso-machining endmills have simple geometric cross-sections as shown in Figures 4.17 and 4.18. The cutting edges were obtained by intersecting two planes without a helix angle. This endmill of 25 μm diameter with five cutting edges (Figure 4.18b) was used to make 25 μm wide × 25 μm deep channel (upper insert) in aluminum. A typical cut depth was 1 μm when milling slots between 20 and 30 μm wide in the three materials selected.

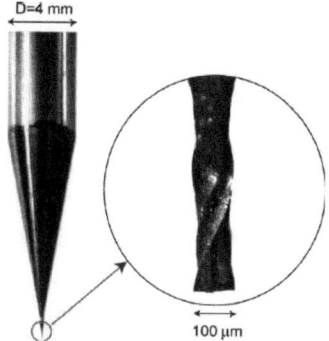

Figure 4.19. *Tungsten carbide micro-endmill with two flutes [FLE 04]*

Commercially available tungsten carbide micro-endmills with helix flutes produced by grinding can be as small as 50 μm in diameter. Figure 4.19 illustrates a typical two-fluted micro-endmill of 100 μm in diameter at the end of a conical tool tip. Other types of micromills are selected in Figure 4.9.

4.4.3. *Machine tools for micromilling*

Micromilling often uses high-speed machining technique with spindle rotations between 40,000 and 50,000 rpm and specialized CAM programming. However, many problems arise when typical micromachining applies to endmill sizes less than 3.3 mm (0.13") diameter and most machined parts will fit inside a 1" (25.4 mm) cube [MCC 06]. Along with a rigid machine, the spindle must be stable to minimize tool vibration and prevent thermal expansion. This is because any vibration or runout at the tool tip deteriorates the surface finish and part accuracy, and can also fracture small endmills. It is well known that small vibrations are amplified relative to tool diameter as it is reduced. For example, a vibration of 0.0001" (0.00254 mm) is a much larger fraction of a 0.010" (3 mm) endmill (1% of the diameter) than a 0.5" (13 mm) endmill (0.02% of the diameter) [MCC 06]. One solution is a direct-change type spindle. In fact, by eliminating a tool holder total run-out caused by tool holder variation is substantially reduced. Usually, a spindle run-out of less than 4 μm is required.

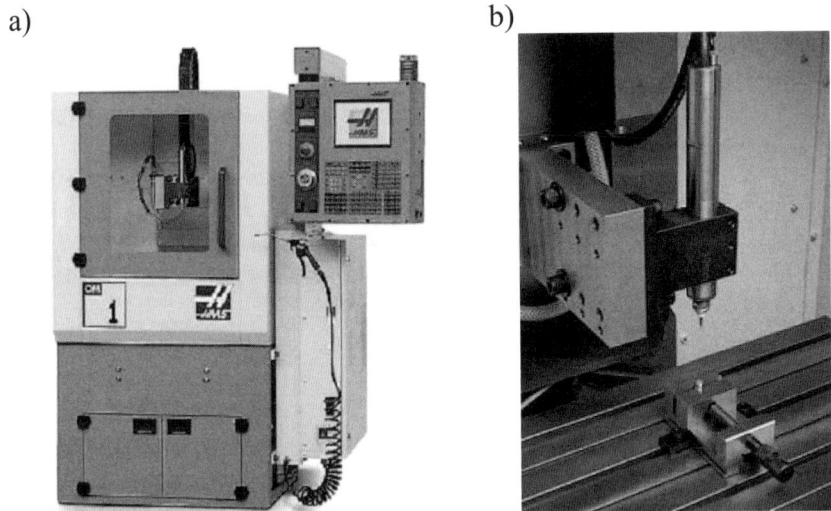

Figure 4.20. *(a) CNC micro-milling machine and (b) view of high-speed spindle [HAY 07]*

As an example, Figure 4.20 shows a CNC micro-milling machine (termed an office mill [HAY 07]) equipped with a 50,000 rpm brushless electric micro-motor spindle with 270 W power rate. Maximum values of feed rate and cutting speed are equal to 19.2 m/min (757 ipm) and 12.7 m/min (500 ipm). It is possible to perform 4-axis machining using a micro-rotary table or 5-axis machining by installing a micro-trunnion table. All possible routes along X, Y and Z axes are equal to 8" (203 mm). Moreover, on this basis small work cells for 3, 4 or 5 axis parts can be created. Positioning accuracy of ±5 μm (±0.0002") and repeatability of ±3 μm (±0.0001") can be achieved. An ultra-compact CNC mill was primarily designed for high production or rapid prototyping of small, precision 2D and 3D parts. The control system is equipped with a 40 GB hard disk drive and Ethernet interface.

The high-speed vertical machining center shown in Figure 4.21, called "smart machine", was designed for automation with a special attention to minimizing vibration and thermal variations. Optionally, the spindle speeds are 30,000 min^{-1}, 42,000 m^{-1} or 54,000 m^{-1}. The machine shown in Figure 4.21 has a table chuck with a palletizing system and a seven-position pallet changer as standard features. The modified bridge design of the machine allows the operator to access the machine from the front and enables the use of a pallet-changing device from the rear. The machining center can also be served by a robot in place of the pallet changer.

Figure 4.21. *High-speed machining center for micromachining [MIK 07]*

The VMC was equipped with several hardware and software modules, consequently called "smart machine" modules, that improve process reliability under high-speed machining regimes. One of these modules, the advanced process system (APS), has a built-in vibration sensor in the spindle together with a spindle diagnostic module. With the help of this system, vibrations which occur during a milling process can be displayed on the monitor of the CNC control as a "G-load" (measuring range of G-loads between 0 and 10 G). Another unique module, labeled intelligent thermal control (ITC), addresses the issue of thermal growth by means of a built-in intelligence on the thermal behavior of the machine. As a result, it automatically compensates for any thermal drift in all operating conditions. Moreover, a warm-up phase or pre-heat cycle is no longer necessary and machining can be carried out with a high level of confidence in a fully unmanned mode. The third module, called a remote notification system (RNS), was installed to obtain information about the operating state of the machining center and the process itself, regardless of the user location. If, for example, the operator has categorized in advance an NC message as important, the plain text information on the generation of unmanned operation is sent to the user on his mobile phone by SMS (short message service).

4.5. Product quality in micromachining

4.5.1. *Quality challenges in micromachining*

In general, micromachining and microtooling/molds have developed a new group of critical tolerance-based processes. It is obviously known that in the dynamically developing medical market, medical devices, electronics and biopharmaceutical manufacturers need new products that create tinier, less invasive, fluid-induced and/or space-saving micro-devices. The tiniest parts must be of high quality to eliminate problems in assembly. Many challenges exist in micromolding systems which include new metrology/inspection and validation techniques, properly sized machines, standardization and part handling [ARO 07]. Inspection techniques for measuring very small micromolded parts require customized vises, tweezers and fixturing. In addition, a low vibration and low noise environment is needed to capture micromolded images accurately. It is also common in macro-components, and specifically with medical devices, to insist on 1.33 C_{pk} or better with respect to performance in drawing dimensions or tolerance. The 1.33 C_{pk} for 0.0001" (0.003 mm) tolerances requires an exceptional accuracy of the gauge used and high precision action of the operator. Component manufacturers and micromolders require similar inspection machines with identical fixtures to validate tolerances in micro components. Customized micro-sized equipment is necessary for drying, handling, weighing, analyzing, quantifying and controlling micro-processes. In the context of this chapter, the most important challenge for microcutting is to produce precise, burr-free holes and slots.

4.5.2. *Burr formation in micromachining operations*

Two important goals in microcutting, which influence the surface finish, are the avoidance of burr formation which is a real "productivity killer" and a good projection of the tool geometry into the material [WUE 01]. Burr formation causes dimensional distortion on the part edge, presents challenges to assembly and handling in sensitive locations on the workpiece and the damage done to the work subsurface from the deformation associated with the exit of the cutting edge [DOR 05]. It is assessed that 2-8% of the manufacturing cost for a major assembly component in automotive industry results from removal of burrs (deburring) or prior changing of tools due to over-ridging [BER 05]. Moreover, deburring operations cost American companies billions of dollars annually, and it is a production headache that frequently causes bottlenecks in production and refinishing [SHO 07]. In order to effectively address burr prevention, the entire "process chain" from design to manufacture must be considered to integrate all the elements affecting burrs, from the part design, including material selection, to the micromachining process [DOR 05]. When implementing burr-minimized production or metal-cutting operations, two main factors must be taken into account, namely: first, the use of burr-minimizing tools, and second, the implementation of optimal process parameters.

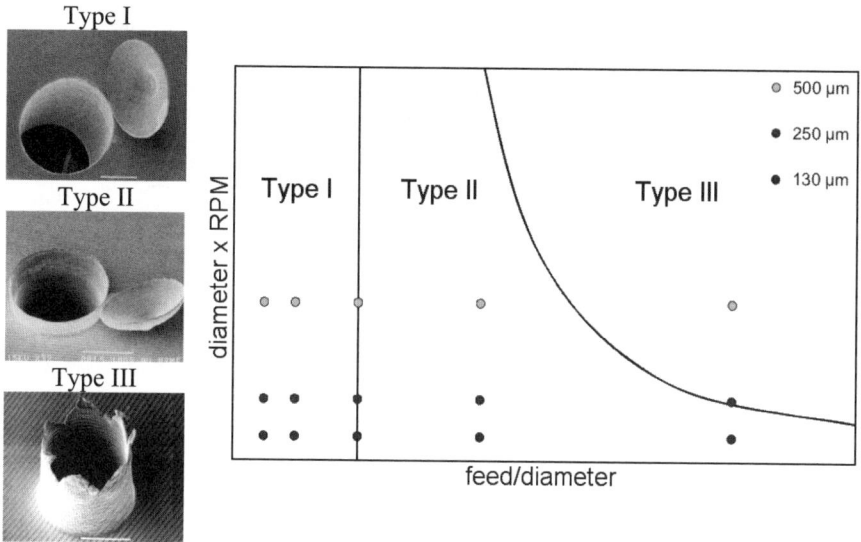

Figure 4.22. *Three typical burrs in microdrilling stainless steel (left) and drilling burr control chart (right) [DOR 05]*

Figure 4.22 illustrates three typical burrs produced in micro-drilling operations on stainless steel parts: small uniform exit burr with cap (type I), large uniform (very thin) exit burr with cap (type II) and crown burr (type III). In general, burr size depends on several cutting conditions including cutting speed and feed rate. It is well confirmed in the drilling burr control chart (right diagram in Figure 4.22) that the type of burr produced and burr height depend on the feed rate and cutting (rotational) speed. For example, crown burrs of 500 μm in height are produced for maximum values of both these cutting parameters.

Another approach to minimizing burr formation in microdrilling operations involves the ultrasonic vibratory technique described in section 4.3.3, which modifies material deformation/fracture conditions during the drilling process. As can be seen in Figure 4.23, microholes of 10 μm diameter produced with axial ultrasonic vibration are free of burrs and cracks. It should be noted that the holes were made in the ductile-regime cutting on very hard glass pieces.

94 Nano and Micromachining

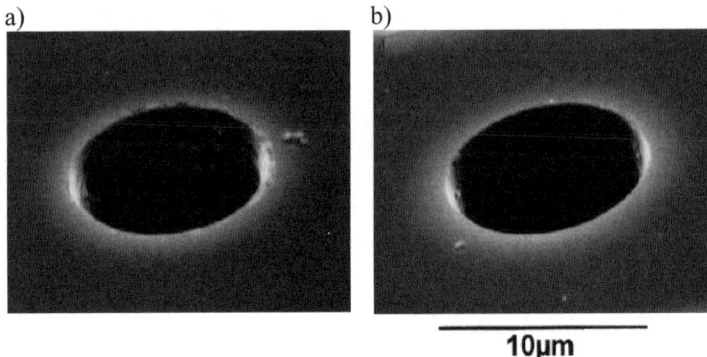

Figure 4.23. *Microholes with a diameter of 10μm and depth of 20 μm drilled (a) without and (b) with vibration (penetration rate=0.05 μm/s) [EGA 02b]*

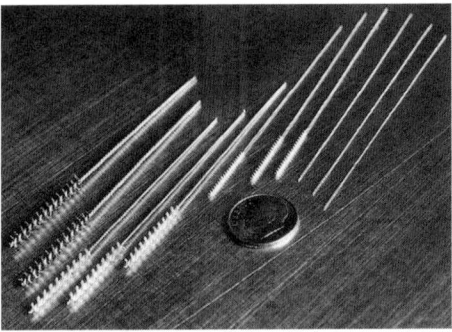

Figure 4.24. *Miniature brushed and twisted wire tools for deburring operations [SHO 07]*

Figure 4.24 presents a range of ball-style miniature brushes, which are highly specialized abrasive tools for specialized surfacing including deburring, edge-blending, plateau honing and deglazing of critical metal surfaces. They contain small abrasive globules that are mounted to flexible filaments using special Flex-Hone technology courtesy of Brush Research Manufacturing (BRM) [SHO 07].

As reported earlier, micromilling with tungsten carbide tools is a proven and effective process for producing complex structures such as molds and microfluidics which are characterized by high aspect ratios (see Table 4.2) and high geometrical complexity but typically rather moderate demands on surface quality. Yet, in many applications the surface quality is vital for the proper functioning of micro-components. For example, molds for micro-injection molding process needs surface roughness better than Rz=1 μm in order to guarantee removing the part from the mold cavity [WEU 01]. Some practical observations are collected when the conventional milling process is downscaled to micrometer dimensions. In particular, the tendency is that the surface quality is better for harder materials but burrs occur more frequently in the hardened state of steel components. Moreover, the groove walls slope away in the hardened state due to high residual stresses which, in turn, are released during the cutting process. In case of microslot in an Al6061-T6 aluminum alloy using the micro mills fabricated by micro-EDM, severe deformation of the 5 μm thin wall occurred, causing the lack of flow within the micro-fluidic system [CHER 07]. This unacceptable result is explained by unexpected vibration and/or lateral bending caused by cutting forces. Another problem occurring in micromilling of micro-thin-walled structures concerns undesirable burrs which are usually produced in conventional milling processes. Basically four types of burrs: (a) primary burr, (b) needle-like burr, (c) feathery burr and (d) minor burr, were produced in micro-milling of micro grooves of the width d ranging from 1 to 3 μm, as shown in Figure 4.25. It is observed in these experiments that the primary burr, the feathery burr and the needle-like burr are produced on the side of micro-slot where up-milling occurs. On the other hand, the burrs tend to be removed with the chip when down-milling is performed.

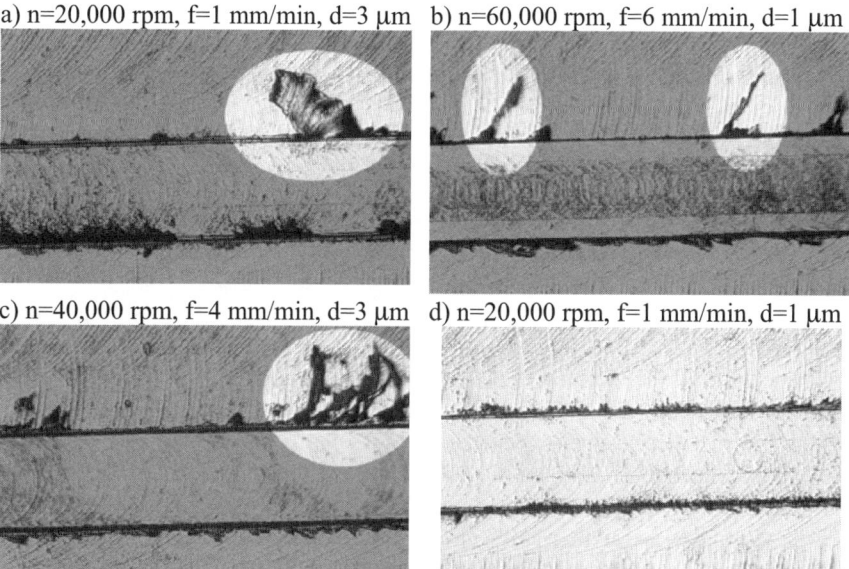

Figure 4.25. *Photos of burrs (100×) in micromachining: (a) primary burr, (b) needle-like burr, (c) feathery burr, (d) minor burr. After [CHE 07]*

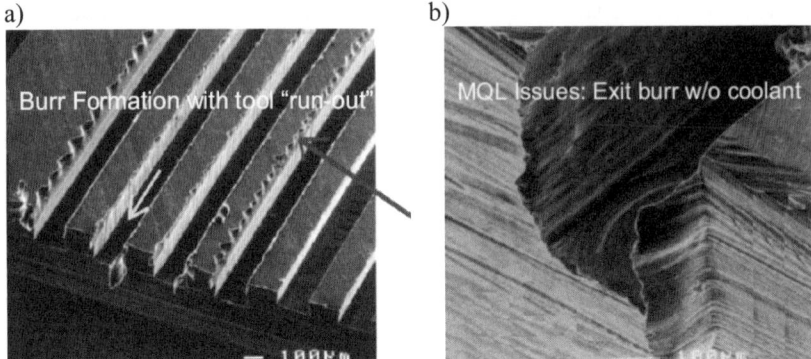

Figure 4.26. *(a) Typical burrs in slot milling of miniature heat exchanger and (b) magnification of side exit burr [DOR 05]*

Similarly to burr photos from Figure 4.25, big wavy burrs remain on the edges of the miniature heat exchanger, as shown in Figure 4.26. Also, big exit burrs were created on the front surfaces as depicted by the SEM image in Figure 4.26b. In general, preventing or eliminating burrs is a very complex technological task which needs integration of many activities such as careful choice of tools, machining parameters, and tool path or special work material and part design [DOR 05], [LIT 06]. In fact, most burrs can be prevented or minimized with appropriate process control.

4.5.3. *Surface quality inspection of micromachining products*

In the micromachining world, appropriate meso- and micro-measuring systems are necessary because part features are often too small to use a conventional contact profilometer or laser. Recently, a special laser system that can simultaneously measure six degrees of freedom using a meso-scale laser head was developed to perform error measurement on the mMT platform [ARO 04]. Similarly, conventional machine tool positioning devices are not the best alternative at this size scale and a new solution employs a device based on voice-coil technology, typically used for high-speed scanning applications that include tissue scanning, where fast acceleration and high positioning resolution are required. For such applications, a piezoelectric-based actuation system can also be suitable. As reported, optical microscope for the observation of micro-cutting process when very tiny tools are cut, i.e. endmills of about 20 μm in diameter and turning tools of about 10 μm wide (machining under the microscope), and electron microscope inspection of machined parts can be an integrated part of the micro-machine tools [ZEL 07]. Inspection techniques in measuring very small micro-parts require customized vises, tweezers and fixturing. Measuring of micro-components is very important for the assembly of functional devices. Due to the miniature size of micro-components, the human eye alone is not sufficient to observe micro-component defects or micro-system performance. Consequently, sophisticated optical systems are required to examine the accuracy and surface finish of such final components. Generally, the smallest touch-trigger probe tip for tactile probing is 0.3 mm diameter with a stylus length in the 2-3 mm range. However, touch-trigger probes in combination with optical image processing sensors, analog scanning probes, and lasers have created a whole class of versatile multi-sensor machines that can be configured for both shopfloor and quality-control laboratory measurement and inspection [LOR 05].

Figure 4.27. *CNC shop floor multi-sensor coordinate measuring system [WER 07]*

For instance, Figure 4.27 shows a shop floor multi-sensor coordinate measuring system with automatic image analysis and CNC positioning. Specific design features are: fixed bridge, moving table configuration, multi-sensor CMM; rigid granite construction; high accuracy ball bearing guideways, and linear measuring scales with low thermal coefficient of expansion. Measuring range: X=400 to 1,500 mm; Y=400 to 1,000 mm; Z=150 to 300 mm. Available measuring sensors include: non-contact image processing (IP) video sensor with fixed telecentric objective; zoom sensor; laser sensor; 3D-Patch, contact-Renishaw point-to-point touch-trigger and dynamic scanning sensors; fiber optic (micro) sensors and contour sensors. Applications cover such parts as plastic connectors and housings; stamped/formed parts; printed circuit boards (PCBs); plastic injection molding components; small dies; medical components, tools and fixtures. In particular, fiber probe sensors feature a tip with a diameter as small as 25 µm, long stylus length (often in excess of 50 mm) and extremely low gauging force (typically in the µN regime) for non-destructive measurements of optical-quality surfaces, deep bores and extraordinary small holes and elements. In addition, the fiber probe tip can be illuminated to measure blind or closed features, counterbores and other geometries/profiles that cannot be measured by transmitted light. Typical applications of this sensor include, for example, tiny turbine blade cooling holes, microscopic-diameter diesel fuel injector spray holes, and infinitesimal cranio-facial surgical screws, often 0.3 mm in diameter or less.

The production of 3D shapes by micromachining makes their testing and handling quite difficult. It should be noted that flat surfaces on steel parts can be more easily measured using rare non-contact with contact devices than, for instance, complex 3D surfaces generated on molded components. Figure 4.28 shows an extended inspection station consisting of an Olympus BX-41 optical microscope with a Prior H-101 motorized X, Y, Z stage, a data processing PC, and a Lumenera LU125 CMOS mono-chromatic camera to capture high–resolution images.

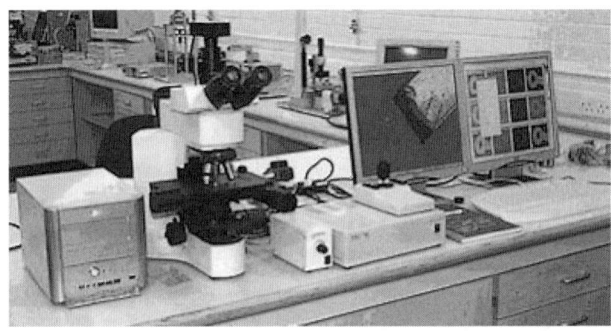

Figure 4.28. *Inspection station for micromolded components [BIB 06]*

Component manufacturers and micromolders require an inspection machine with identical fixtures to validate near-micron tolerance in micro-components. In addition, extremely clean HEPA-filtered air-controlled rooms are necessary for repeatability or measurements. It is standard, for micro-sized parts, to insist on a process capability of 1.33 C_{pk} or better with respect to the performance of appropriate devices.

4.6. References

[ARO 04] ARONSON R.B., "Micromanufacturing is growing", report on micromanufacturing, SME, www.sme.org, 2004, p. M1-M18.

[ARO 07] ARONSON R.B., "Micromanufacturing; the new frontier", *Manufacturing Engineering*, vol. 138, no. 5, 2007, p. 89-100.

[BER 05] BERGER K., "Burr minimization along the process chain-activities in a German automotive industry group", *Proc. 8th CIRP Int. Workshop on Modeling of Machining Operations*, Chemnitz, 10-11 May, 2005, p. 45- 48.

[BIB 06] BIBBER D., "Is going micro worth the effort?", *Manufacturing Engineering*, vol. 137, no. 6, 2006, p. 115-123.

[BRI 01] BRINKSMEIER E., BLÄBE R., "Advances in precision machining of steel", *Annals of the CIRP*, vol. 50, no. 1, 2001, p. 385-388.

[BYR 03] BYRNE G., DORNFELD D., DENKENA B., "Advancing cutting technology", *Annals of the CIRP*, vol. 52, no. 2, 2003, p. 483-507.

[CHA 06] CHAE J., PARK S.S., FREIHEIT T., "Investigations of micro-cutting operations", *Int, J. Machine Tools & Manufacture*, vol. 46, no. 3-4, 2006, p. 313-332.

[CHE 07] CHERN G-L., WU E. Y-J., CHENG J-CH., JAO J-CH., "Study on burr formation in micro-machining using micro-tools fabricated by micro-EDM", *Precision Engineering*, vol. 31, no. x, 2007, p. 122-129.

[CIT 07] "High-speed, high-precision machining of tiny precision parts", report by Citizen Cincom, www.marucit.com/products/R07.html.

[DOR 94] DORF R.C., KUSIAK A., *Handbook of Design, Manufacturing and Automation*, New York, John Wiley & Sons, 1994.

[DOR 05] DORNFELD D., "Micromachining and burr formation for precision components" *Proc. MTTRF*, San Francisco, July 2005, p. 66-80.

[EGA 02a] EGASHIRA K., MIZUTANI K., "Micro-drilling of monocrystalline silicon using a cutting tool", *Precision Engineering*, vol. 26, no. 3, 2002, p. 263-268.

[EGA 02b] EGASHIRA K., MIZUTANI K., "Ultrasonic vibration drilling of microholes in glass", *CIRP Annals, Manufacturing Technology*, vol. 51, no. 1, 2002, p. 339-342.

[FLE 04] FLEISCHER J., MASUZAWA T., SCHMIDT J., KNOLL W., "New applications for micro-EDM", *J. Mater. Proc. Technol*, vol. 149, no. 1-3, 2004, p. 246-249.

[GRZ 08] GRZESIK W., *Advanced Machining Processes of Metallic Materials*, Amsterdam, Elsevier, 2008.

[HAY 07] HAYES D., "Office machine review", Hass Intl. Sales Training, www.haas.com.

[HOR 08] "Horn-leaders in grooving technology", www.hornusa.com.

[KEN 07] "KM Micro Quick-Change Tooling", www.kennametal.com.

[KUD 07] KUDŁA L., "Microcutting techniques for micromachining", *APE'2007*, Warsaw, 13-16 June 2007, Publ. House of the Institute for Sustainable Technologies, Radom, p.281-290.

[LIT 06] LITWINSKI K. M., MIN S., LEE D-E., DORNFELD D., LEE N., "Scalability of tool path planning to micro machining", *Proc. 1st Conference on Micromanufacturing (ICOMM)*, University of Illinois, Urbana-Champaign, 13-15 September, 2006, paper no. 28.

[LOR 05] LORINICZ J., "Measuring micro parts", *Manufacturing Engineering*, vol. 135, no. 5, 2005, p. 77-84.

[MAI 08] "Fast and versatile Swiss turning centers", *Automatic machining*, February, 2008, p. 32-33 (www. maier-swiss.com).

[MAS 00a] MASUZAWA T., "State of the art of micromachining", *Annals of the CIRP*, vol. 49, no. 2, 2000, p. 473-488.

[MAS 00b] MASUZAWA T., TÖNSHOFF H.K, "Three-dimensional micromachining by machine tools", *Annals of the CIRP*, vol. 46, no. 2, 1997, p. 621-628.

[MCC 06] McCORNIK M., DeBOER CH., "Micromachining techniques, fixturing and end mill selection in high-precision VMC parts", *Machining Technology*, no. 3, 2006, www.sme.org.

[MIK 07] "Smart machine – a new dimension in milling technology", www.mikron-ac.com.

[MIN 08] Minitool Micro Drilling Systems, www.minitoolinc.com.

[RAH 07] RAHMAN M., LIM H.S., NEO K.S., KUMAR A.S., WONG Y.S., LI X.P., "Tool-based nanofinishing and micromachining", *J. Mater. Proc. Technol*, vol. 185, no. 1-3, 2007, p. 313-332.

[SAN 07] "Small part machining", www.coromant.sandvik.com.

[SHO 07] "Hone smooths rough auto edges", Shop solutions, *Manufacturing Engineering*, vol. 138, no. 5, 2007, p. 188.

[WER 07] www.werth.de (Werth Messtechnik GmbH, Giessen, Germany).

[WUE 01] WUELE H., HÜNTRUP H., TRITSCHER H, "Micro-cutting of steel to meet new requirements of miniaturization", *Annals of the CIRP*, vol. 50, no.1, 2001, p. 61-64.

[ZEL 07] ZELINSKI P., "Machining under the microscope", www.msonline.com/articles/010504.

Chapter 5

Microgrinding and Ultra-precision Processes

5.1. Introduction

Microgrinding and ultra-precision processes such as nanogrinding are an aspect of advanced manufacturing that has been growing rapidly in recent years. For instance, lenses and mirrors are manufactured to precise and ultra-precise standards. Most of these applications require a crack-free surface. Generally, hard and brittle materials such as glass, silicon and germanium are commonly used for making these products. Traditionally, optical glasses require grinding and finishing by a polishing process in order to remove the damage caused by the previous operation and to obtain a flat surface. The conventional ground surface of glass results in a fine finish due to brittle fracture during the removal process and this surface needs to be polished to 20 nm R_{max} to reduce absorption and scattering of light on the glass surface. However, advances in precision machining of brittle materials have led to the discovery of a ductile regime in which material removal is performed by plastic deformation. Fracture mechanics predicts that even brittle solids can be machined by the action of plastic flow, as is the case in metal, leaving crack-free surfaces when the removal process is performed at less than a critical cut depth. This means that under certain controlled conditions, it is possible to machine brittle materials such as ceramics and glass using single-point diamond tools so that material is removed by plastic flow, leaving a smooth and crack-free surface [YAN 02], [SCH 91], [KOM 96], [NAM 93], [PUT 89].

In order to overcome the problems of excessive tool wear, multi-point cutting (grinding) has become economical, especially when machining hard and brittle

---

Chapter written by Mark J. JACKSON and Michael D. WHITFIELD.

materials [MOR 95], [IKA 91], [NAM 99], [OHM 95], [BLA 90]. Ultra-precision surface grinding with electrolytic dressing (ELID) provides in-process dressing of the wheel achieving almost 100% ductile regime machining with a finish surface without the need for subsequent polishing when grinding optical glasses and silicon-based materials.

Several models have been suggested to explain ductile mode theory in machining processes. A critical cut depth and a feed rate concept for ultra-precision machining has been proposed, as shown in Figure 5.1 [MOR 95]. An initial model was developed based on indentation fracture mechanics analysis [MOR 95]. According to Scattergood and colleagues, fracture initiation plays a central role for ductile-regime machining. A critical penetration depth, d_c, for fracture initiation is

$$d_c = \beta \left[\frac{K_c}{H}\right]^2 \left[\frac{E}{H}\right] \qquad [5.1]$$

where Kc is fracture toughness, H is hardness and E is elastic modulus. β is a factor that will depend upon geometry and process conditions such as tool rake angle and coolant.

A round nosed diamond tool moves through the workpiece, as shown in Figure 5.1. Figure 5.1 shows a projection of the tool perpendicular to the cutting direction. Using the critical depth concept, fracture damage will initiate at the effective cutting depth d_c ($t_c \cong d_c$ as shown in Figure 5.1) and will propagate to an average depth y_c as shown. If the damage does not continue below the cut surface plane, ductile regime conditions are achieved. The cross feed f determines the position of d_c along the tool nose. Larger values of f move d_c closer to the tool. It is important to note that when ductile regime conditions are achieved, material removal still occurs by fracture. The model proposed was verified by interrupted tests and the following relationship was obtained:

$$\frac{Z_{eff}^2 - f^2}{R^2} = \frac{d_c^2}{f^2} - 2\left[\frac{d_c + y_c}{R}\right] \qquad [5.2]$$

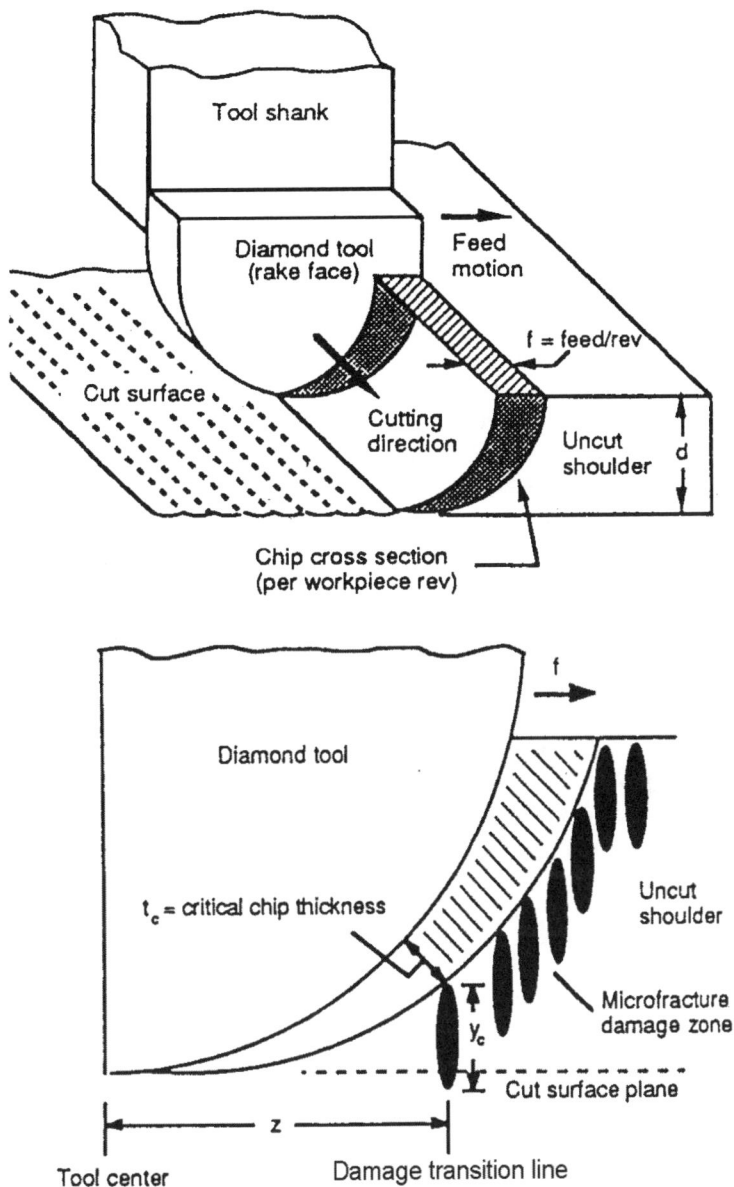

Figure 5.1. *Scattergood's model of ultra-precision machining showing a 3D view of a diamond tool while cutting material and a cross-sectional view of the tool and workpiece [MOR 95]*

where R is the tool nose radius and the other parameters are defined as shown in Figure 5.1. When using [100] Ge single crystal, it was found that dc = 130 nm and yc = 1,300 nm when using equation [5.2] with a tool having a nose radius of 3.175 mm and -30° rake angle.

5.2. Micro and nanogrinding

In order to achieve fully ductile mode grinding, material removal must take place at the nanoscale. The nanogrinding process is a process that relies on using a nickel-coated ceramic material with microscale diamond particles bonded to it that are cubo-octahedral in shape to machine nanoscale features in a variety of workpiece materials. The diamonds are bonded to the piezoelectric material by gaseous deposition, laser cladding or directly bonding a porous tool to the material via an adhesive paste. The process is executed by applying a known sinusoidal frequency to the piezoelectric crystal in order to achieve a desired oscillatory displacement. Rapid vibration of the crystal will allow material removal rates to be increased, thus making it a nanomanufacturing process. The nanogrinding process is accompanied by wear of the diamond grains, and the rate of this wear plays an important role in determining the efficiency of the nanogrinding process and the quality of the nanomachined surface. Wear mechanisms in nanogrinding processes appear to be similar to that of single-point cutting tools, the only difference being the size of swarf particles generated.

Abrasive grains with blunted cutting edges (wear flats) and abrasive grains with sharp cutting edges are released from the surface of the piezoelectric crystal before they have a chance to grind nanoscale chips from the surface of the workpiece. The process suffers with a loss of diamond grains even when the interfacial adhesion between diamond and piezoelectric material is very good. A more closely related process that has been reported widely is that of the wear of probes used in atomic force microscopy. However, these observations were purely experimental and after no explanation of how to design probes that inhibit, or retard, wear.

A performance index used to characterize diamond wear resistance is the grinding ratio, or G-ratio, and is expressed as the ratio of the change in volume of the workpiece removed, Δv_w, to the change in the volume of the diamond abrasive grain removed, Δv_s; it is shown in equation [5.3]:

$$G = \Delta v_w / \Delta v_s \qquad [5.3]$$

Grinding ratios for processes at the nanoscale have not yet been characterized. However, the complexities of wear for abrasive materials at any scale lead us to believe that the variety of different and interacting wear mechanisms involved, namely, plastic flow of abrasive, abrasive crumbling, chemical wear, etc., makes diamond wear at the nanoscale too complicated to be explained using a single theoretical model [JAC 01].

5.2.1. *Nanogrinding apparatus*

The experimental apparatus consisted of holding a polished specimen between the jaws of a vise so that a piezoelectric crystal oscillator traverses back-and-forth across the specimen, thus machining the specimen by creating a cut depth between the diamonds adhered on the piezoelectric crystal and the workpiece material. Workpiece materials were polished with a 100 nm sized polishing compound. All samples were divided into four sections and each section was analyzed prior to machining and after machining occurred. The workpieces were mounted in a vise that was attached to a x-y-z linear slide in order to achieve accurate positioning of the workpiece. The piezoelectric crystal was mounted on a steel framework that was orthogonal to the workpiece. The whole unit was located within a tetrahedral space frame to dampen excess vibrations.

5.2.2. *Nanogrinding procedures*

When the crystal and workpiece were aligned, the cut depth was incremented in stages of 10 nm. The motion of the diamonds attached to the piezoelectric crystal generates a machining effect that is caused by the action of diamonds grinding into the workpiece material. Tracks, or trenches, are created by the diamond grain gouging the surface of the material when an electric current is applied. The material is removed until the end of the oscillating motion creates the material to plough. The oscillation mechanism can be described as a restricted bending mode that simulates a shear displacement of the crystal. At this point it is normal procedure to explain how to estimate the number of grains contacting the surface of the workpiece.

The estimation of the number of active cutting grains is made quite simply by driving the diamond coated piezoelectric ceramic at the prescribed specific metal removal rate into a piece of lead. The impression that the grinding wheel produces in the length of lead is equal to the number of cutting points that are active during the grinding stroke at that particular cut depth.

The motion of the diamonds imparted by oscillating the crystal in the bending mode that causes a shear displacement to occur, which contributes to ploughing of the material at the end of the nanogrinding stroke, is shown in Figure 5.2.

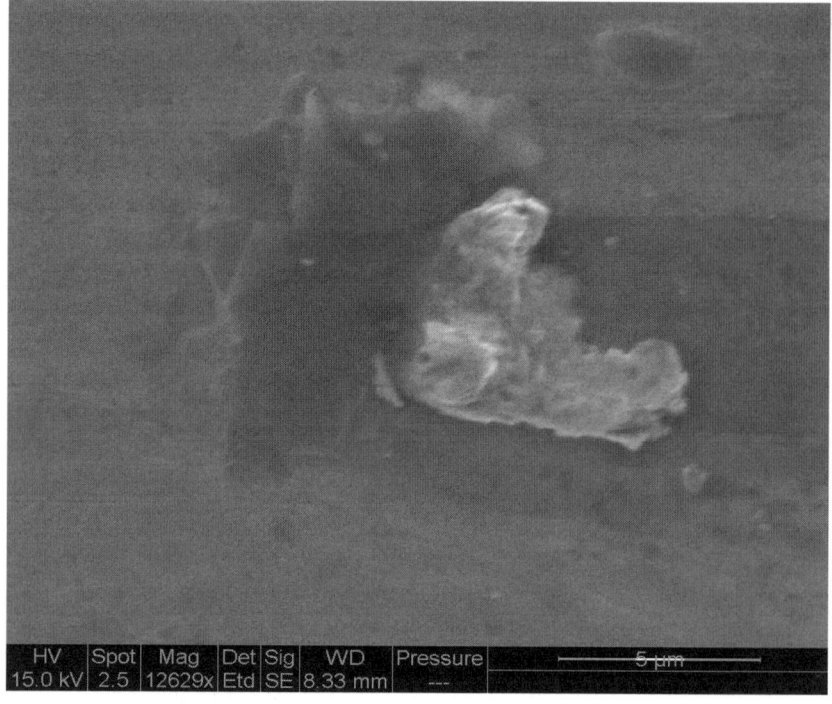

Figure 5.2. *Ploughed material at the end of a nanomachined track*

The effect of using the piezoelectric to machine tracks, or trenches, in engineering materials opens up the prospect of nanomanufacturing products that require geometric features such as channels so that fluids and mixed phase flows can be manipulated in devices such as micro- and nanofluidic "lab-on-a-chip" products.

The measured force components of the nanogrinding operation are measured using a dynamometer. These force components are then applied to a model of abrasive grain by dividing the grinding force data into the number of active cutting grains over an area that simulates the abrasive grain-workpiece contact area. Stresses established in this area are calculated using finite elements. The wear of the piezoelectric material by diamond loss, expressed in terms of a grinding ratio, and its relationship to the stress levels established in the model grain has been investigated using a stress analysis method.

5.3. Nanogrinding tools

Porous nanogrinding tools are composed of abrasive particles (submicron size) embedded in a vitrified bond with porosity interspersed between grinding grains and bonding bridges. The porosity level is approximately 15-21%. Figure 5.3 shows the image of a nanogrinding tool prior to laser modification. The vitrified bonds are specially engineered to promote the formation of texture that creates ridges of cutting planes and nanogrinding "peaks" of α -Al_2O_3 in the preferred (012), (104), and (110) planes. The peaks created due to laser modification of the surface aid the nanogrinding process. Initial studies are focused on using vitrified structures.

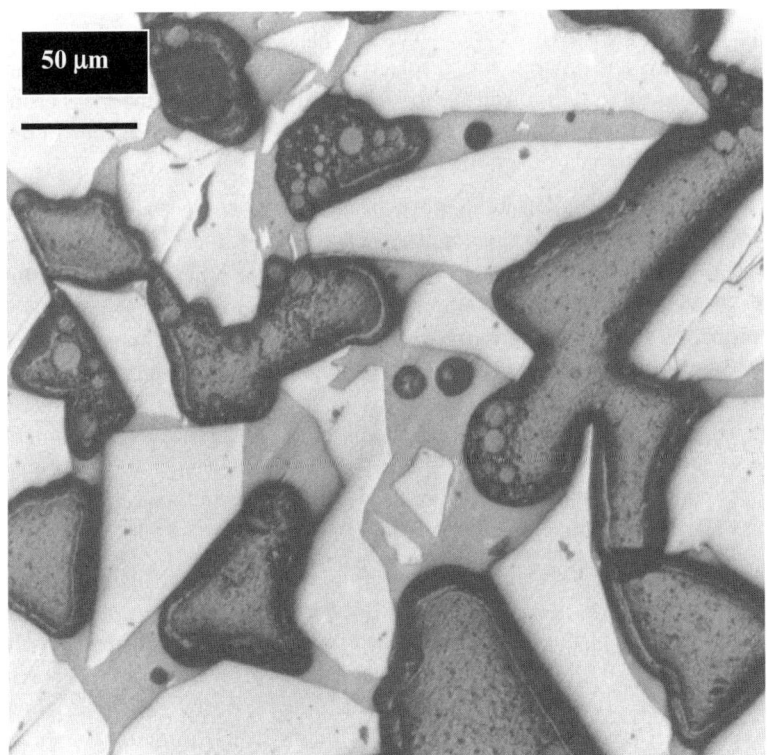

Figure 5.3. *Structure of the porous tool used for nanogrinding*

Vitrified bonds are composed of glasses formed when clays, ground glass frits, mineral fluxes such as feldspars and chemical fluxes such as borax melt when the grinding wheel is fired at temperatures in the range of 1,000°C to 1,200°C. With reference to raw material nomenclature, a "frit" is a pre-ground glass with a predetermined oxide content, and a "flux" is a low melting point siliceous clay that reduces surface tension at the bond bridge-abrasive grain interface. Again, a "pre-fritted" bond is a bond that contains no clay minerals (i.e. clays and fluxes) and "firing" refers to vitrification heat treatment that consolidates the individual bond constituents together [LUN 59]. Considering individual bond constituents, mineral fluxes and ground glass frits have little direct effect on the ability to manufacture grinding wheels. However, most clays develop some plasticity in the presence of water (from the binder), which improves the ability to mold the mixture so that the wheel, in its green state, can be mechanically handled.

Clays and clay-based fluxes contain an amount of free quartz that has a detrimental effect on the development of strength during vitrification heat treatment. Clays are used to provide vitrified grinding wheels with green strength during the heat treatment process. However, when the glass material solidifies around the particles of clay and quartz, the displacive transformation of quartz during the cooling stage of vitrification leads to the formation of cracks in the glass around the quartz particle. The strength of the bonding bridge is impaired and leads to the early release of the abrasive particle during the metal cutting.

The basic wear mechanisms that affect vitrified grinding wheels are concerned with grain fracture during metal cutting, fracture of bond bridges, mechanical fracture of abrasive grains due to spalling, and fracture at the interface between abrasive grain and bond bridge. Failure in vitrified silicon carbide grinding wheels is more probable due to the lack of a well-developed bonding layer between the abrasive grain and glass bond bridge. The bonding layer is approximately a few micrometers in thickness, and is caused by the use of a high clay content bonding system. High glass content bonding systems tend to aggressively decompose the surface of silicon carbide abrasive grains. In vitrified corundum grinding wheels, high glass content bonding systems are used extensively and lead to bonding layers in excess of 100 micrometers in thickness.

In addition to the formation of very thin bonding layers in vitrified silicon carbide grinding wheels, the use of high clay content bonding systems implies that there is an increase in the amount of quartz in the bond bridges between abrasive grains. Although the likelihood of decomposition on silicon carbide surfaces is reduced, the probability of bond bridge failure is increased due to the increased quartz content. Therefore, the dissolution of quartz is of paramount importance in order to compensate for thinner interfacial bonding layers. The dissolution of quartz in a liquid phase does not require a nucleation step. One process that determines the rate of the overall reaction is the phase-boundary reaction rate, which is fixed by the movement of ions across the interface. However, reaction at the phase boundary leads to an increased concentration at the interface. Ions must diffuse away from the reaction interface so that the reaction can continue. The rate of material transfer and the diffusion rate are controlled by molecular diffusion in the presence of a high-viscosity liquid phase. For a stationary solid in an unstirred liquid, or in a liquid with no fluid flow produced by hydrodynamic instabilities, the dissolution rate is governed by molecular diffusion.

The effective diffusion length over which mass is transported is proportional to $\sqrt{Dt}$, where D is the diffusion coefficient and t is time, and therefore the change in thickness of the solid, which is proportional to the mass dissolved, and varies with $\sqrt{t}$. Natural, or free, convection occurs because of hydrodynamic instabilities in the liquid which gives rise to fluid flow over the solid. This enhances dissolution kinetics. Generally, a partially submerged solid undergoes more dissolution near to the solid-liquid interface. Below this interface the dissolution kinetics of the solid can be analyzed using the principles of free convection.

The boundary layer thickness is determined by the hydrodynamic conditions of fluid flow. Viscous liquids form much thicker boundary layers, which tends to impede material transfer. Higher liquid velocities promote the formation of thinner boundary layers and make more rapid material transfer possible. Considering the dissolution of quartz in glass materials, the high viscosity and slow fluid flows combine to give thick boundary layers. Also, the diffusion rate is much slower in viscous silicate liquids than in aqueous solutions, thus giving a tendency for the reaction process to be controlled by material transfer phenomena rather than by interface reactions.

Difficulties encountered when developing a dissolution model arise from the fact that the phase boundary between quartz particle and molten glass moves during the diffusion process. The problem of a fixed boundary can be solved without difficulty, although this is not equivalent to the conditions associated with a moving boundary between the quartz particle and a highly viscous glass melt. The development of dissolution models is required to determine the magnitude of quartz remaining in the bonding system after a period of heat treatment. The models are then compared with experimentally determined quartz content of the bonding systems using x-ray diffraction techniques.

5.3.1. *Dissolution modeling*

When densification occurs in a vitrified grinding wheel, the cooling rate is reduced to prevent thermal stress cracking in the bonding layer between abrasive particles. Cooling rates are reduced when crystalline inversions involving volume changes occur. The inversion range for quartz and cristobalite are 550°C–580°C and 200°C–300°C, respectively. Since the formation of cristobalite is rare in most vitrified bonding systems used for grinding wheels, the rapid displacive transformation of quartz tends to promote crack formation in bonding bridges. Once the grinding grain is lost the remaining bonding bridges can be modified using a high power laser to create an oriented texture that forms "peaks" of α-Al_2O_3 in the preferred (012), (104), and (110) planes.

When quartz-containing bonds begin to cool from the soaking, or vitrification, temperature it is thought that the liquid phase relieves stresses resulting from the thermal expansion mismatch between itself and the phases, β-quartz, β-cristobalite and mullite, to at least 800°C. At 800°C, stresses will develop in quartz particles and the matrix that causes micro-cracking to occur. The shrinkage behavior of quartz and the glass phase has been described by Storch *et al.* [STO 84]. Between the temperature range, 573°C and 800°C, the glass phase shrinks more than the quartz phase that causes tangential tensile stresses to form cracks in the matrix. At 573°C, β-quartz transforms to α-quartz, which causes residual stresses to produce circumferential cracking around quartz particles. Some of these cracks have been seen to propagate into the glass phase [BIN 62]. Similar observations occur in the cristobalite phase. Spontaneous quartz cracking has been found to occur over a temperature range that depends on the size of the quartz particles [FOR 51]. Particles with a diameter larger than 600 micrometers cracked spontaneously at 640°C, whereas smaller particles with a diameter of less than 40 μm cracked at 573°C. This observation agrees with temperature dependent micro-cracking reported by Kirchhoff *et al.* [KIR 82]. To maintain the integrity of the bond bridges containing coarse quartz particles, the grinding wheel must remain at the vitrification temperature until the quartz particles have dissolved.

The dissolution model assumes that at a constant absolute temperature, T, a particle of quartz melts in the surrounding viscous glass melt, and that the rate of change in the volume of quartz present in the melt at a particular instant in time is proportional to the residual volume of quartz. The above assumption is based on the fact that alkali ions diffuse from the viscous glass melt to the boundary of the quartz particle, thus producing a dissolution rim around each quartz particle. A high reaction rate will initially occur which continuously decreases as the quartz particle is converted to a viscous melt.

Jackson and Mills [JAC 97] derived a mathematical relationship that accounts for the change in density when β-quartz transforms to α-quartz on cooling from the vitrification temperature, thus,

$$m_{T,t} = M\gamma \exp\left(-At^{1/2} \exp\left[\frac{-B}{T}\right]\right) \quad [5.4]$$

where $m_{T,t}$ is the residual mass fraction of quartz at a constant time and temperature couple, M is the original mass fraction of quartz prior to heat treatment, γ is the density ratio of β-quartz and α-quartz, A and B are constants, t is time, and T is absolute temperature. The model was compared with experimental data determined using the powder x-ray diffraction method.

5.3.2. *Preparation of nanogrinding wheels*

The raw materials used in the experimental study were Hymod Prima ball clay, standard porcelain china clay, potash feldspar and synthetic quartz (supplied as silica flour). The chemical analysis of the raw materials is shown in Table 5.1. Rational analysis of the raw materials was performed to reveal the mineralogical composition of the raw materials. The rational analysis appears in Table 5.2. The bond mixture described is one typically used in vitrified silicon carbide grinding wheels where the erosion of the abrasive grain is reduced using high clay content bonding systems.

Oxide (wt.%)	China clay	Ball clay	Potash feldspar	Quartz
Al_2O_3	37	31	18.01	0.65
SiO_2	48	52	66.6	98.4
K_2O	1.65	1.8	11.01	0.35
Na_2O	0.1	0.2	3.2	0.04
CaO	0.07	0.2	0.09	0.00
MgO	0.03	0.3	0.09	0.00
TiO_2	0.02	0.9	0.00	0.07
Fe_2O_3	0.68	1.1	0.11	0.03
Loss on ignition	12.5	16.5	0.89	0.20

Table 5.1. *Chemical analyses of raw materials*

Fusible bonding systems using a mixture of ball clay and potassium-rich feldspar were made to test the model developed by Jackson and Mills [JAC 97]. The ball clay used contained 12.77 wt.% quartz, and the feldspar contained 4.93 wt.% quartz. The bonding system was composed of 66 wt.% ball clay and 34% feldspar. The initial quartz content, M, of the bond mixture was 10.1 wt.%. The bond mixture described is one typically used in high-performance vitrified corundum grinding wheels.

Compound (wt. %)	China clay	Ball clay	Potash feldspar	Quartz
Quartz	4.05	12.77	4.93	98.40
Orthoclase	0.00	15.23	64.96	0.00
Kaolinite	79.70	62.71	2.17	0.00
Mica	13.94	0.00	0.00	0.00
Soda feldspar	0.8	1.69	27.07	0.00
Miscellaneous oxides/losses	1.51	7.60	0.87	1.60

Table 5.2. *Mineralogical analyses of raw materials*

The raw materials were mixed in a mortar, pressed in a mold and fired at various temperatures. A heating rate of 2.9°C min^{-1} was employed until the vitrification temperature was reached. The typical soaking temperature was varied between 1,200°C and 1,400°C for "sintering" bond compositions, and 950°C and 1,050°C for "fusible" bond compositions in order to simulate industrial firing conditions. The samples were cooled at a rate of 1.8°C min^{-1} to avoid thermal stress fracture. The fired samples were crushed to form a fine powder in preparation for x-ray diffraction.

5.3.3. Bonding systems

The dissolution model was compared with experimental data using the x-ray powder diffraction method. X-ray diffraction of the raw materials was performed on a Phillips 1710 x-ray generator with a 40 kV tube voltage and a 30 mA current. Monochromatic Cu kα radiation, λ = 0.154060 nm, was employed. A scanning speed of 2° per minute for diffraction angles of 2θ was used between 2θ angles of 10° and 80°, and the x-ray intensity was recorded using a computer. The spectrum was then analyzed and compared with known spectra.

Powder specimens were prepared by crushing in a pestle and mortar in preparation for quantitative x-ray diffraction. To eliminate the requirement of knowing mass absorption coefficients of ceramic samples for quantitative x-ray diffraction, Alexander and Klug [ALE 48] introduced the use of an internal standard. First, the ceramic sample is crushed to form a powder – the sizes of particles should be small enough to make extinction and micro-absorption effects negligible. Second, the internal standard to be added should have a mass absorption coefficient at a radiation wavelength such that intensity peaks from the phase(s) being measured are not diminished or amplified. It should be noted that the powder diffraction mixture should be homogenous on a scale of size smaller than the amount of material exposed to the x-ray beam, and should be free from preferred orientation. The powder bed that is subjected to "x-rays" should be deep enough to give maximum diffracted intensity.

The expected equilibrium phases from the fired mixtures are quartz (unreacted and partially dissolved), mullite, cristobalite and glass. However, from the samples tested, the compounds quartz, mullite and glass were successfully detected. A calibration curve was constructed using a suitable internal standard (calcium fluoride), a diluent (glass made by melting potash feldspar) and a synthetic form of the phase(s) to be measured. Synthetic mullite had a purity greater than 99.8%, whilst powdered quartz had a purity greater than 99.84% SiO_2. The method used for quantitative analysis of ceramic powders was developed by Khandelwal and Cook [KHA 70].

The internal standard gave a fairly intense (111) reflection (d = 0.1354 nm) lying between the (100) reflection for quartz (d = 0.4257 nm) and the (200) reflection for mullite (d = 0.3773 nm). Using copper kα radiation (λ = 0.15405 nm), the corresponding values of diffraction angle 2θ are: (100) quartz = 20.82°; (111) calcium fluoride = 28.3°; and (200) mullite = 32.26°. Mass fractions of the crystalline phases in the mixture can be read from the calibration lines by measuring the intensity ratio of the phase(s) to the internal standard. In order to calculate the mass fractions of quartz and mullite in the mixture, the height of the chosen diffraction peak and its width at half-height were measured from the diffraction spectrum. The product of these two measures was then compared with that of the internal standard, and the resultant intensity ratio was used to find the exact mass fraction of the phase(s) measured in the glass that has been subjected to x-rays.

5.3.4. Vitrified bonding systems

In addition to comparing the experimental results to the dissolution model, results published in the literature were also used to test the accuracy of the model.

The composition of the experimental mixtures was matched to those specified by Lundin [LUN 59]. Lundin's experimental mixtures were composed of 25 wt.% quartz (13.2 μm particle size), 50 wt.% clay (kaolin), and 25 wt.% flux (potassium feldspar - 25 μm particle size).

The constants A and B for the sintering bonding system were calculated:

$$A = 5.62 \times 10^8 \qquad [5.5]$$

$$B = 33374 \qquad [5.6]$$

from which the experimental activation energy, Q, is 132.65 k cal/mole. The residual quartz content for the sintering bonding system is

$$m_{T,t} = 26.25 \cdot \exp\left[-5.62 \times 10^8 \cdot t^{1/2} \cdot e^{\frac{-33374}{T}}\right] \qquad [5.7]$$

The data comparing Lundin's experimental results, the author's experimental results and the dissolution model are shown in Table 5.3.

When the data are plotted as the logarithm of ($-\ln[m/M]/t^{1/2}$) versus the reciprocal of absolute temperature, 1/T, then all data fits a straight line relationship. The gradient was calculated to be 33,374, the constant B using two data points. Lundin's experimental gradient gave a value of 32,962 using the least squares method, and 34,000 for the present work. The corresponding activation energies for both systems are 131 kcal/mole for Lundin's work [LUN 59] and 135 kcal/mole for the present work, respectively. Figures 5.4 and 5.5 show the effects of time on residual quartz content at different temperatures together with comparative experimental data.

A comparison was made with dissolution models published in the literature. One of the earliest models was derived by Jander [JAN 27]. The equation can be expressed as:

$$\left(1 - \sqrt[3]{1-Z}\right)^2 = \left\{\frac{C_1 \cdot D}{r^2}\right\} \cdot t \qquad [5.8]$$

where Z is the volume of quartz that has been dissolved, r is the original particle radius, and D is the diffusion coefficient for the diffusing species. This equation can be transformed into mass fractions using Archimedes' law, thus:

$$\left(1 - \sqrt[3]{\frac{m}{M}}\right)^2 = C_2 \cdot t \qquad [5.9]$$

where C is a constant dependent on soaking temperature and initial particle size of quartz. Krause and Keetman [KRA 36] expressed the dissolution of quartz as a function of isothermal firing time, viz:

$$M - m = C_3 . \ln t \qquad [5.10]$$

where M is the initial quartz content and m is the residual quartz content after time, t. The unit of time here is seconds such that after one second of firing, the residual quartz content is equal to the initial quartz content. Monshi's dissolution model [MON 90] can be transformed into the following equation assuming isothermal firing conditions:

$$\ln\left\{\frac{m}{M}\right\} = -C_6 \sqrt{t} \qquad [5.11]$$

Jackson and Mills' model [JAC 97] for isothermal firing conditions is transformed into:

$$\ln\left\{\frac{m}{\gamma.M}\right\} = -C_7 \sqrt{t} \qquad [5.12]$$

where γ is the density ratio of $\beta-$ and $\alpha-$ quartz. Constants for all the equations presented here are calculated using quartz mass fraction data after an 18 hour firing. The constants are dimensioned in seconds. The equations shown were compared with experimental data generated by Lundin [LUN 59] for a clay-based material containing 40 wt.% kaolin, 40 wt.% quartz, and 20 wt.% feldspar.

Temp. (°C)	Time (hrs.)	Lundin's exp. result (wt.%)	Exp. result (wt.%)	Jackson and Mills' [JAC 97] result (wt.%)
1200 (1473K)	1	24.1	24.2	24.2
1200	1	24.7	24.3	24.2
1200	1	26.1	24.8	24.2
1200	2	23.7	23.8	23.4
1200	2	23.6	23.9	23.4
1200*	2	23.4	23.4	23.4
1200	4	21.3	22.2	22.3
1200	8	20.3	20.9	20.8
1200	18	19.0	18.5	18.6
1200	18	18.9	18.6	18.6
1200	48	15.2	15.1	14.9
1250 (1523K)	1	22.7	22	22.1
1250*	2	20.6	20.6	20.6
1250	4	18	18.5	18.6
1250	8	15.5	16	16.2
1250	18	12.6	12.5	12.6
1250	48	8.3	7.8	8.0
1300 (1573K)	0.5	22.6	20.4	20.6
1300	0.5	21	20.9	20.6
1300	1	20	18.3	18.6
1300	2	16.1	15.9	16.2
1300	4	13.4	12.8	13.2
1300	8	10	9.7	9.9
1300	18	5.9	5.8	6.1
1300	50	1.6	1.8	2.3
1300	120	0.3	0.2	0.6

Table 5.3. *Residual quartz content of a sintering bonding system at various vitrification temperatures. Lundin's [LUN 59] experimental data are compared with the authors' experimental data and the model [JAC 97]. The asterisk indicates values used for deriving the constants used in the theoretical model*

According to the transformed equations, the mass fraction of quartz can be calculated as follows:

– Jander's model [JAN 27]

$$m = 41.9 \cdot \left(1 - \left\{1.55 \times 10^{-6} \cdot t\right\}\right)^{3/2} \qquad [5.13]$$

– *Krause and Keetman's model* [KRA 36]

$$m = 41.9 - (2.58 \cdot \ln t) \quad [5.14]$$

– *Monshi's model* [MON 90]

$$m = 41.9 \cdot e^{-4.5 \times 10^{-3} \sqrt{t}} \quad [5.15]$$

– *Jackson and Mills' model* [JAC 97]

$$m = 41.73 \cdot e^{-4.5 \times 10^{-3} \sqrt{t}} \quad [5.16]$$

The transformed equations are then tested using data provided by Lundin [LUN 59]. Referring to Table 5.3, it can be shown that the mass fraction of quartz obtained using the equations derived by Jander [JAN 27], and Krause and Keetman [KRA 36] did not agree with Lundin's experimental results [LUN 59].

Time (hrs.)	Lundin's exp. data [LUN 59]	Jander [JAN 27]	Krause and Keetman [KRA 36]	Monshi [MON 90]	Jackson and Mills [JAC 97]
0	41.9	41.9	0.00	41.9	41.9
0.5	35.9	41.72	22.55	34.61	34.76
1	32.8	41.54	20.76	31.97	32.12
2	29.2	41.19	18.97	28.58	28.72
4	23.2	40.49	17.18	24.39	24.51
8	19.5	39.11	15.39	19.49	19.59
18	13.3	35.72	13.30	13.30	13.36
24	10.7	33.74	12.56	11.13	11.19
48	6.9	26.18	10.77	6.43	6.51
120	3.6	7.85	8.96	2.17	2.17
190	2.7	0.00	7.22	1.00	1.01
258	2.0	0.00	6.43	0.54	0.55

Table 5.4. *Residual quartz content for different soaking times at 1,300°C for a sintering bonding system composed of 40 wt.% kaolin, 40 wt.% quartz and 20 wt.% feldspar (Lundin's mixture number M21 [LUN 59]) compared with other dissolution models*

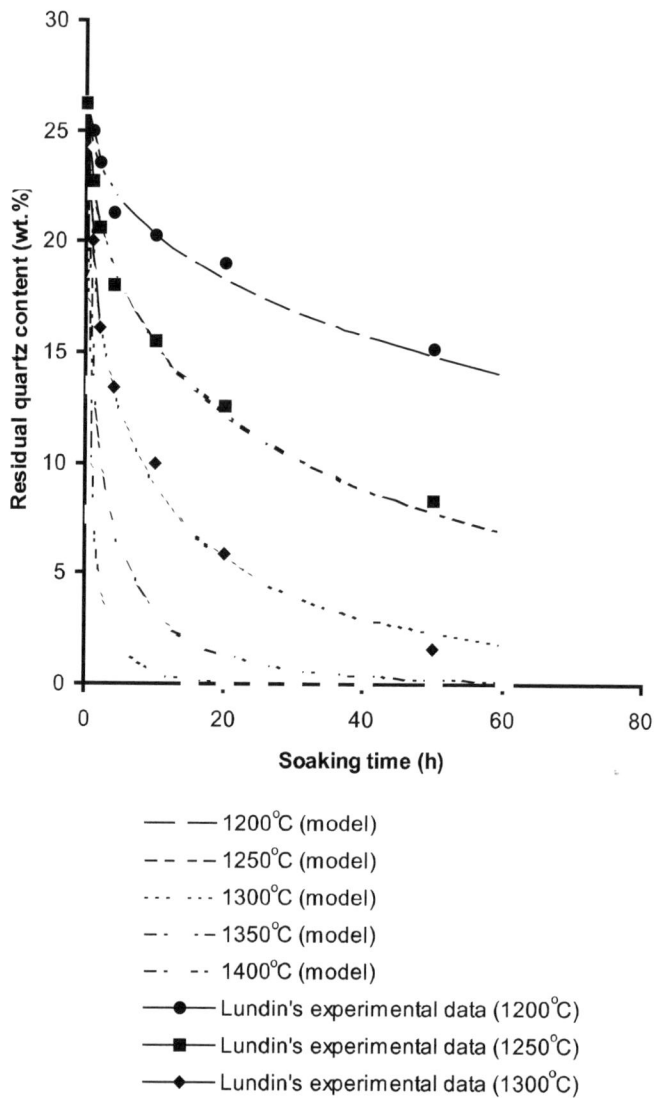

Figure 5.4. *Effect of time on residual quartz content of a sintering bonding system according to Jackson and Mills' model [JAC 97] and compared with Lundin's experimental data [LUN 59]*

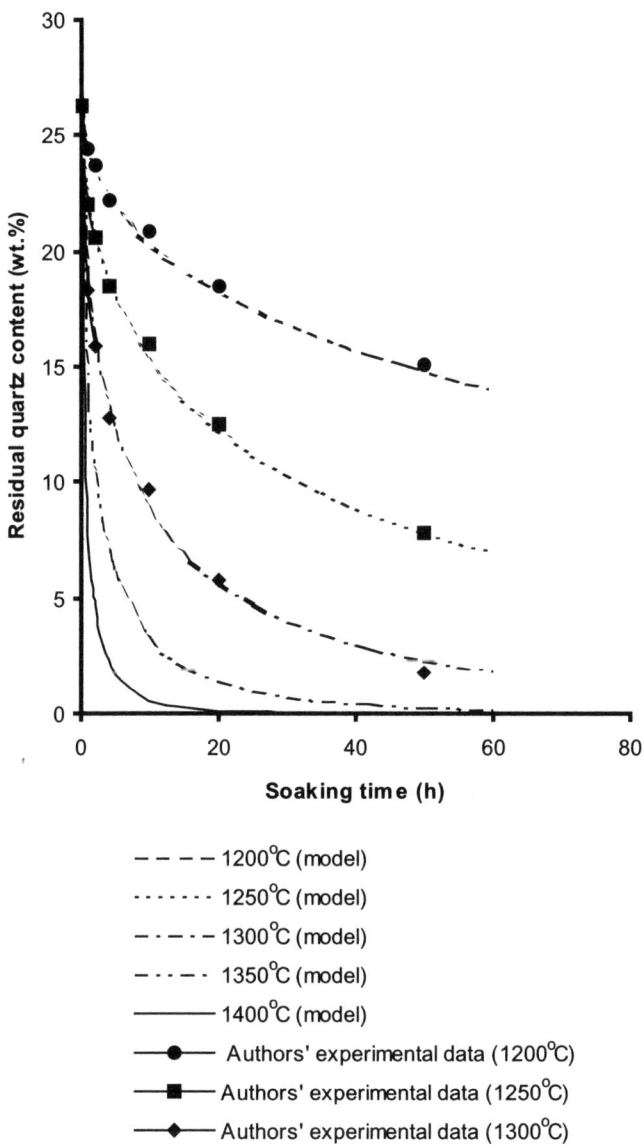

Figure 5.5. *Effect of time on residual quartz content of a sintering bonding system according to Jackson and Mills' model [JAC 97] and compared with the authors' experimental data*

The results obtained using Monshi's model [MON 90] are in much better agreement compared to Lundin's data. However, the results obtained using Jackson and Mills' model [JAC 97] are more accurate at predicting the mass fraction of quartz remaining owing to the differences in the density of quartz. After long periods of heat treatment, the model predicts lower magnitudes of mass fractions for quartz when compared to Lundin's experimental results [LUN 59].

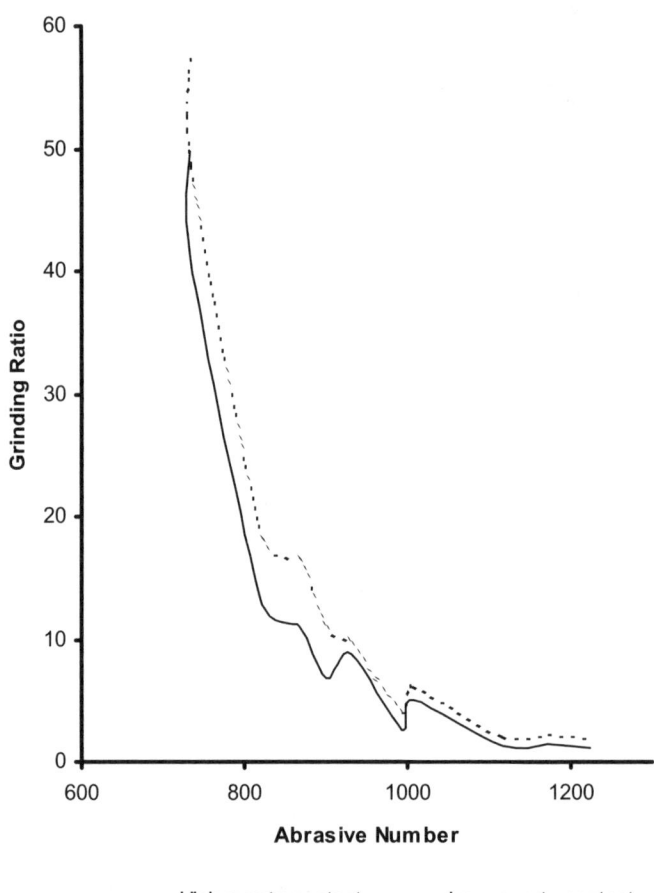

Figure 5.6. *Effect of the abrasive number on the grinding ratio for a high-quartz content and a low quartz-content bonding system*

Figure 5.6 shows the effect of using a high and a low quartz content bonding system on the wear of vitrified corundum grinding wheels grinding a large number of tool steel materials [JAC 95]. The classification of tool steels is in the form of an abrasive hardness number, which is a weighted average of the number of carbides contained within the tool material. As shown in Figure 5.6, the grinding ratio, or G-ratio, is a measure of the efficiency of the grinding wheel. It is the quotient of the volume of workpiece material removed and the volume of the wheel material removed. The figure demonstrates the effectiveness of reducing the quartz content of the bonding system of porous nanogrinding tools.

5.4. Conclusions

For perfectly sharp diamond-coated piezoelectric ceramic materials, grain fracture appears to be the dominant cause of abrasive material loss during a grinding operation, especially when grinding engineering materials. Grain fracture is much more likely to be caused by mechanically induced tensile stresses within abrasive grains than by mechanically induced compressive stresses. The best indicator of diamond grain performance during a nanogrinding operation under different operating conditions is the level of tensile stress established in abrasive grains. Nanogrinding processes are especially suited to machining optical materials such as glass, germanium, zinc sulfide, etc.

High tensile stresses are associated with grain fracture and low grinding ratios in perfectly sharp diamond-coated piezoelectric ceramic materials. Finite element models of perfectly sharp grinding grains can be applied to the nanogrinding process where the dominant wear mechanism is grain fracture. Nanogrinding is a process that has demonstrated its suitability for development into a nanomanufacturing process of the future. The dissolution model derived by Jackson and Mills has been compared with experimental data using sintering vitrified bonding systems that are used extensively with high performance nanogrinding tools. The results predicted by the model compare well with the experimental results presented in this chapter. However, over longer periods of isothermal vitrification, the model becomes less accurate due to the assumptions made in the dissolution model. The model may be of use when predicting the mass fraction of quartz using high temperature firing cycles that are characterized by short soaking periods.

The development of porous nanogrinding tools lies in their ability to be dressed using a directed photon beam that sharpens worn grains or promotes the formation of textured peaks from the vitreous bonding system. The development of quartz-free bonding systems with a high corundum content is of paramount importance if porous nanogrinding tools are to be effective when machining engineering materials at the nanoscale.

5.5. References

[ALE 48] ALEXANDER, I.E., KLUG, H.P., "X-ray diffraction procedures", *Anal. Chem.*, Vol. 20, 1948, p. 886.

[BIN 62] BINNS, D.B., "Some physical properties of two-phase crystal-glass solids", in G.H. Stewart (ed.), *Science of Ceramics,*, Vol. 1, Academic Press, New York, 1962, p. 315-334.

[BLA 90] BLAKE, P.N., and SCATTERGOOD, R.O., "Ductile regime machining of germanium and silicon", *Journal of the American Ceramic Society*, Vol. 73, no. 4, 1990, p. 949-957.

[FOR 51] FORD, W.F., and WHITE, J., "Effect of heat on ceramic whiteware bodies", *Trans. J. Brit. Ceram. Soc.*, Vol. 50, 1951, p. 461.

[IKA 91] IKAWA, N. *et al.*, "Ultraprecision metal cutting – the past, the present and the future", *Annals of the CIRP*, Vol. 40, no. 2, 1991, p. 587-594.

[JAC 95] JACKSON, M.J., A study of vitreous-bonded abrasive materials, PhD Thesis, Liverpool John Moores University, UK, December 1995.

[JAC 97] JACKSON, M.J. and MILLS, B., "Dissolution of quartz in vitrified ceramic materials", *J. Mat. Sci.*, Vol. 32, 1997, p. 5295 - 5304.

[JAC 01] JACKSON, M.J., "Vitrification heat treatment during the manufacture of corundum grinding wheels", *Journal of Manufacturing Processes*, Vol. 3, 2001, p. 17-28.

[JAN 27] JANDER, W., Reaktion im festen zustande bei hoheren temperaturen (Reactions in solids at high temperature), *Z. Anorg. U. Allgem. Chem.*, Vol. 163, 1927, p. 1-30.

[KHA 70] KHANDELWAL, S.K., and COOK, R.L., "Effect of alumina additions on crystalline constituents and fired properties of electrical porcelain", *Amer. Ceram. Soc. Bull.*, Vol. 49, 1970, p. 522 - 526.

[KIR 82] KIRCHOFF, G. *et al.*, "Spontaneous cracking of quartz investigated using acoustic emission techniques",, *J. Mat. Sci.*, Vol. 17, 1982, p. 2809.

[KOM 96] KOMANDURI, R., "On material removal mechanisms in finishing of advanced ceramics and glasses", *Annals of the CIRP*, Vol. 45, no. 1, 1996, p. 509-513.

[KRA 36] KRAUSE, P., KEETAN, E., Zur kenntnis der keramischen brennvorgange (On combustion processes in ceramics), *Sprechsaal*, Vol. 69, 1936, p. 45-47.

[LUN 59] LUNDIN, S.T., *Studies on Triaxial Whiteware Bodies* (Almqvist and Wiksell), Stockholm, Sweden, 1959.

[MON 90] MONSHI, A., Investigation into the strength of whiteware bodies, PhD Thesis, University of Sheffield, UK, 1990.

[MOR 95] MORRIS, J.C. et al., "Origins of the ductile regime in single-point diamond turning of semiconductors," *Journal of the American Ceramic Society*, Vol. 78, no. 8, 1995, p. 2015-2020.

[NAM 93] NAMBA, Y., and ABE, M. "Ultraprecision grinding of optical glasses to produce super-smooth surfaces", *Annals of the CIRP*, Vol. 42, no. 1, 1993, p. 417-420.

[NAM 99] NAMBA, T. et al., "Ultraprecision surface grinding of chemical vapor deposited silicon carbide for X-Ray mirrors using resinoid-bonded diamond wheels", *Annals of the CIRP*, Vol. 48, no. 1, 199, p. 277-280.

[OHM 95] OHMORI, H. and NAKAGAWA, T., "Analysis of mirror surface generation of hard and brittle materials by ELID grinding with superfine grain metallic bond wheels", *Annals of the CIRP*, Vol. 44, no. 1, 1995, p. 287-290.

[PUT 89] PUTTICK, K.E. et al., "Single-point diamond machining of glasses", *Proc. Royal. Society, London*, Vol. A426, 1989, p. 19-30.

[SCH 91] SCHINKER, M.G., "Subsurface damage mechanisms at high-speed ductile machining of optical glasses", *Precision Engineering*, Vol. 13, no. 3, 1991, p. 208-218.

[STO 84] STORCH, W. et al., "Investigation of cracking in whiteware ceramics", *Berichte Deut. Keram. Ges.*, Vol. 61, 1984, p. 325.

[YAN 02] YAN, J. et al. "Ductile regime turning at large tool feed", *Journal of Materials Processing Technology*, Vol. 121, 2002, p. 363–372.

Chapter 6

Non-Conventional Processes: Laser Micromachining

6.1. Introduction

Laser micromachining is based on the interaction of light with solid matter. As a result of the complex interaction between light and matter, small amounts of material can be removed from the surface. Two different phenomena are in operation, namely: pyrolithic processing and photolithic processing. Pyrolithic processing is composed of heating, melting and ablating the material at the surface, and photolithic processing is based on the direct breaking of chemical bonds in a wide range of materials.

A laser is a very intense beam of light that removes material by breaking atomic bonds. Light can be described as energy packets, or photons, that have a wavelength and a frequency. Monochromatic light is special because it is composed of only one wavelength, with the same phase, and is coherent compared to incandescent light, which is composed of different wavelengths. Optical energy is transferred to the electrons by absorption, which essentially increases the energy of electrons by increasing the electron vibration which is sensed as heat. This chapter describes the basic principle of using lasers to fabricate features at the microscale.

Chapter written by Grant M. ROBINSON and Mark J. JACKSON.

6.2. Fundamentals of lasers

6.2.1. *Stimulated emission*

Atoms can exist only at defined energy levels; there are no transition states. The pumping mechanism raises the energy level of an atom, and after a short time the atom tries to return to its original state of energy. To do this it ejects the energy in the form of a photon. It is the combined effect of producing many photons that produces laser light. This rise and decay of energy states to produce a photon can take place in a number of mediums, which emit different wavelength photons resulting in different laser types for different applications.

If the temperature of the cavity is too high, the lower energy level cannot be transferred to its ground state fast enough, thus ceasing the production of laser light. As such, the maximum temperature that can be sustained by the system defines the maximum power of the laser. Carbon dioxide lasers can employ different cooling strategies. Slow flow lasers achieve cooling through the cavity walls. A uniform gain is achieved across and along the cavity giving it a good mode and making it particularly well suited for laser cutting. Fast axial flow lasers achieve cooling by convection of the gas through the discharge zone. The gas enters cold and leaves hot, usually at a speed of around 300–500 m/s. The configuration of the system produces a symmetric power distribution of the beam. The cavity length is such that it produces a low Fresnel number and therefore the beam has a low order and can easily be focused to a small point. The gain can be up to 500 W/m. The power these lasers produce is proportional to the cross-sectional area; therefore, producing short lasers would seem like a good way to keep the design compact. However, such designs produce high mode numbers, making the beam difficult to focus easily. There are other types of laser that work on the same principle of gaining photons from the energy decay of various particles.

A solid crystal made up of yttrium, aluminum and garnet with the addition of neodymium ions makes up the Nd:YAG laser. It is therefore known as a solid-state laser. It works on a similar principle to the CO_2 laser; high energy levels are achieved by the neodymium ions, and their subsequent loss of energy when returning to the ground state releases a photon. Energy is supplied by a flashlamp, and photons are emitted when the energy level drops from a high state to a low state. The decay time for each energy level is very short. After the 1.06 µm laser radiation has been emitted, a terminal state is reached. In order to reach the ground state, further cooling is required. Because the pumping efficiency is low, a great deal of energy has to be put into the system, which requires cooling to prevent thermal distortions of the beam. A krypton lamp powers the Nd^{3+} ions in the YAG rod. The Q-switch is a way of controlling the beam: it can be a mechanical chopper, a bleachable dye, an optoelectrical shutter or an acousto-optic switch (a piezoelectric material changes its density when an electric current is applied, which in turn changes the refractive index of the material. Hence, the material acts as an optical grating, therefore controlling the beam).

The Q-switch itself must be cooled because when it is blocking the beam, it absorbs energy. Q switches can be configured to provide pulse rates between 0 and 50 kHz. The idea is that while the shutter is closed, energy builds up and a high peak power is released when the shutter opens, e.g., a 20 W Nd:YAG Q-switched laser can produce 6 ns pulses of 1 mJ/pulse, which is 100 kW. Beam output can be changed by a process known as frequency doubling. Non-linear optical devices can be swamped with photons, they can absorb two or more photons, and therefore rise to a higher energy state. This energy is released in one step; the resulting radiation has half the wavelength and twice the photon energy, e.g., a 1.06 µm Nd:YAG beam can be converted to 0.530 µm green light, and the process can then be repeated again, which would produce ultraviolet light.

A problem with the Nd:YAG laser is the poor efficiency (10-15%) of converting flash lamp energy to high Nd^{3+} energy states, which produces waste heat that can distort the YAG rod leading to poor beam quality (M^2 approximately 15-100). The flash lamp can be replaced by using a diode laser with better efficiency (30-40%). This produces much better values of M^2, as low as 1.1 [STE 98].

6.2.2. *Types of lasers*

Diode lasers are very similar to Nd:YAG lasers where electrical energy is translated via the diode into electron excitation and eventually light is emitted. Usually they are grouped and stacked together to form high power lasers. Semiconductor materials have a bandgap such that if the electrons have enough energy (provided by an electrical field) they convert from a non-conductive state to a conductive state. Part of this change of state can release photons, rather than the laser actions already discussed. GaAs is a common type. It has a bandgap energy of 1.35 eV corresponding to a wavelength of 905 nm. Diode lasers have the advantage of being small and affordable. Diodes tend to emit over a frequency range but they can be tuned by a grating. Low-power diode lasers have low power conversion efficiencies of around 2%, whereas high power diode lasers have high power efficiencies up to 30%. A 5 µm wide strip can produce 100 mW of power, a 50 µm wide strip can produce 0.5 W, and a 500 µm wide strip can produce 4 W. Stacking such arrays can produce even more power, although they usually have high divergence values that make them unsuitable for laser applications.

Excited dimer molecules (hence excimer) decay and release their laser radiation. The excited dimer Kr^+F^- lasts around 5-15 ns and the photons produced are in short pulses of 20 ns that are spread in a 0.4 nm band, which is large, but the power is around 35 MW (0.2 J/pulse).

Ti:sapphire lasers produce pulses in the femtosecond time regime. Short, low-power nanojoule femtosecond pulses are created by an erbium doped fiber laser. Amplification of these low-power pulses is required to produce a useful power output, but there are problems associated with this. These problems are overcome by using chirped pulse amplification (CPA). The pulse is stretched, amplified and then compressed to create a high-intensity femtosecond pulse. The problem with directly amplifying short pulses is their tendency to stretch in time: the pulse may be amplified but it will no longer be on the femtosecond *timescale*. Femtosecond pulses also tend to destroy the medium through which they travel. CPA avoids these problems by the basic principle of stretching short pulses to reduce the peak intensity. The pulse is then compressed back to the original timescale. The femtosecond pulse is made up of several wavelengths. The diffraction gratings or chirping mirrors separate out the different wavelengths of the pulse. This is done because the grating causes the individual frequencies to reflect at different angles. Thus, light from the same pulse travels different distances, which tends to stretch the pulse and reduce the peak intensity. The stretched pulses are let into the regenerative cavity for amplification. The pulse is sent along a certain path. At a point in this path it passes through a Ti:sapphire crystal which is the gain medium, and this is pumped by an Nd:YAG laser. Each time the stretched pulse passes through the cavity it gains a little more energy, receiving a boost from the interactions in the cavity. Along a part of this path there is a component that reflects the pulse if it is not of high enough energy. It is then sent back into the regenerative amplifier to gain more energy until it reaches some critical value. When this value is reached, the pulse encounters the device and it now has enough energy to escape. Thus, the pulses have been amplified in terms of power but are stretched in terms of time. An exact reversal of the process that stretched the pulse is now applied to recapture the original pulse length. The pulses are now at full power and are on the femtosecond timescale. The pockel cell and polarizer perform the tasks of timing the entry of pulses into the regenerative amplifier and determining how the pulses are let out of the cavity and how the gain is received [McG 02].

6.2.3. *Laser optics*

Optics are used to change the beam diameter (increases the power, supplies the beam to a device with a specific inlet diameter, and changes the focal distance to the substrate), the beam direction, or the beam intensity. The conventional laws of optics apply but special care must be taken in terms of coatings and the material the optic is made from. For an ideal beam,

$$\theta W = \lambda / \pi \qquad [6.1]$$

the realistic laser beam equation is:

$$\theta W = M^2 \lambda / \pi \qquad [6.2]$$

where θ = half the divergence angle, ω = beam waist size and λ = wavelength. As everything is ideal here this expression is true over the whole length of the beam. However, in reality perfect conditions never exist. M^2 is a beam quality number, which is one for a perfect beam. The minimum spot size can be found in the following way:

$$\theta_f = D/2f \qquad [6.3]$$

resulting in a minimum spot size

$$\delta = \frac{4}{\pi} M^2 \lambda \frac{f}{D} \qquad [6.4]$$

where f is the focal distance, D is the lens diameter and δ is the minimum spot size diameter. Minimum spot size governs the minimum feature size. Remember that in conventional machining, small features can be created with relatively large cutting tools. Clearly M^2 is an important parameter to control as it one of the major influences of the minimum spot size. Hence, a high-quality beam is desirable for micromachining. For micromachining, a small spot size is required. This is obtained when $M^2 = 1$, with a short wavelength, and a short focal lens length. Examples of beam properties are shown in Table 6.1.

6.2.4. Beam quality

The laser cavity geometry affects the quality of the beam. Cavity length and mirror configuration cause different types of laser spot to be produced. Oscillating light between the mirrors is not forced to travel along the optical axis. It is these oscillations that create transverse electromagnetic modes (TEMs). The number of off-axis modes is described by the Fresnel number, $N = a^2/\lambda L$, where N is the Fresnel number, a is the radius of output aperture, λ is the wavelength, and L is the cavity length. Off-axis oscillations would be visible on a screen placed at mirror B when differences in wavelengths between on-and-off axis oscillations differs by a whole number of wavelengths n when using Pythagoras' theorem.

Laser	Wavelength, λ (μm)	Power, P (W)	w.θ (mm.mrad)	Beam quality, M^2	Spot diameter with f/4 lens, δ (μm)
HeNe	0.63	0.002	0.2	0.98	3
Fine drilling Nd:YAG	1.06	100	6	10	50
Nd:YAG	1.06	1000	25	80	500
Q-switched Nd:YAG	1.06	10	2	3	15
Q-switched Nd:YAG	1.06	100	6	10	50
Q-switched Nd:YAG	1.06	1000	25	80	500
CO_2	10.6	1000	10	1.5	80
Copper vapor	0.51	20	0.5	3	8
Ti:sapphire	0.78	1			
Excimer	0.193-0.351	100	20	200	

Table 6.1. *Properties of laser beams*

The laser cavity geometry affects the quality of the beam. Cavity length and mirror configuration cause different types of laser spot to be produced. These are known as TEMs. Hence,

$$a^2 + L^2 = (L + n\lambda)^2 \quad [6.5]$$

simplifying and ignoring the $n^2\lambda^2$ term because it is very small,

$$a^2 = 2Ln\lambda \quad [6.6]$$

$$n = a^2/(2L\lambda) = \text{(Fresnel number)}/2 \quad [6.7]$$

Low Fresnel numbers yield low order modes; the off-axis oscillations are lost by diffraction and cannot aid in the amplifying process.

6.2.5. Laser-material interactions

The principle of laser material removal is that laser light is focused on the material surface where energy is absorbed. This energy is converted to heat. The maximum depth where absorption occurs is called the penetration depth and leads to the conduction of heat into the material. The energy absorbed for a CO_2 laser is about 20% for a Nd:YAG, or approximately 40-80% for a femtosecond laser; the rest is reflected. Special care must be taken with optics: special anti-reflective coatings must be applied to prevent beams in unwanted areas of the system. Laser energy is converted to heat only as far as the penetration depth. At the penetration depth the power density is approximately 1/e of the original density at the surface. For a CO_2 laser this is about 15 nm, while for an Nd:YAG laser this is about 5 nm. Conduction of heat into the bulk depends on the timescale of the pulses. The mechanism that lasers use to remove material is called ablation. Ablation is achieved by melting and vaporizing the material, which is then ejected from the vicinity of the surface and is governed by power density. However, the ejected material can be deposited near to the melt region, where it freezes and is known as the "re-cast" layer (Figure 6.1).

For a power density of 10^{-9} W/cm^2 the melting point is reached in 300 ns; if the power density is increased 10-fold, this time is reduced 10-fold to 30 ns. The high vaporization rate then causes a shockwave that can reach speeds of 3 km/s. Material is expelled because of high pressure created in the melt and explosive-like boiling of the superheated liquid after the end of the laser pulse. This ejected material can re-form; sometimes it is not always totally vaporized, especially if it is from a deep trench that may cool quickly and reform. Nanosecond lasers produce heat flow whereas femtosecond lasers do not because no melt pool has been produced. Femtosecond laser pulses machine with minimal heat generation where the heat affected zone (HAZ) is given by the equation HAZ $\sim (D.t)^{1/2}$, where D is the thermal diffusion coefficient and t is the pulse duration. Heat flow produces the melt pool, around which the bulk of the material heats up. If rapid heating occurs then the damage to the material is incurred at a greater depth than is desired; this is the HAZ. Clearly the relationship HAZ $\sim (D.t)^{1/2}$ states that shorter pulse durations will decrease the amount of HAZ. The diffusion length for a 20 ns laser is 365 times longer than a 150 fs laser.

Figure 6.1. *Re-cast layer created close to the melt region*

The nanosecond diffusion length is 1 x 10^{-4} m, and the approximate diffusion length for metals is 1 x 10^{-6} m. The femtosecond diffusion length is approximately 4 x 10^{-7} m, where a femtosecond is equal to 10^{-15} tenths of a second. Materials react to induced heat pulses in picoseconds, i.e. 10^{-12} tenths of a second. Therefore, the material can be heated and removed before the surrounding material can react; hence there can be no heat flow in the femtosecond case. Material at the top of the surface is removed by evaporation, and at the sidewalls of the hole or trench material is forced away by the plasma. Both effects cause re-deposition of material elsewhere.

Plasma pressure is exerted on the liquid; at the end of the pulse, the pressure suddenly drops and causes boiling of the superheated liquid to occur. Ablation occurs at the end of the pulse, and it has been discovered that for a given fluence there is a maximum ablation depth for a given pulse length. Material removal rates can be difficult to calculate due to re-deposition of molten material. Ablation of metals is caused by the absorption of laser energy that is a three-step process: (1) absorption of photon by electrons (10^{-15} s); (2) energy transferred to the lattice (10^{-12} s); and (3) heat transferred to the lattice (10^{-12} s). Ablation at the end of the pulse is due to the relatively thick superheated layer, which continues to evaporate as long as the surface is above the boiling temperature. Re-deposition may occur in the path of a track which has yet to be machined. In this case more material has to be removed. Deposition may occur elsewhere in a non-critical area. This is what makes calculating material removal rates difficult. In the case of femtosecond pulses during the solid-to-liquid transformation, there is no melting because it happens so quickly that it is considered instantaneous. Owing to this effect, there is no heat transfer and there is a debate as to whether the HAZ exists or not.

If the HAZ does exist then the heat source could come from the plasma. It is expelled and extinguished rapidly, but a new one is also created just as rapidly. This means that a constant plasma heating source needs to be controlled by scan speed and power, i.e. process parameters. The types of laser discussed thus far produce pulses on the nano, pico and femtosecond scales. Each timeframe has different characteristics for material removal.

6.3. Laser microfabrication

6.3.1. *Nanosecond pulse microfabrication*

When electromagnetic waves interact with the particle on the surface of the material, the electron will re-radiate or be constrained by the lattice; if enough energy is put into the material, the lattice breaks down and the material begins to melt. Further heating causes evaporation and plasma formation to occur. When laser radiation hits a surface it is absorbed, transmitted or reflected depending on the material. The laser radiation that interacts with the particle has a magnetic and electric component. When the radiation passes over a small elastically bound charged particle, the particle is set in motion by the electric field. This induced force is so small it cannot affect the nucleus but can affect the electrons. If the electrons were left to vibrate there would be no net gain in its energy, e.g. its motion would be in the form of a sinusoidal wave, with a few positive and a few negative motions resulting in zero energy change. However, if the electrons are involved in a collision, the path will be upset and they will gain some energy. If the radiation is at a lower potential than the ionization energy for the particles, then no absorption occurs.

Usually there are free or conduction electrons available for this purpose, which are called seed electrons. Once the seed electron gains enough energy, then further collisions will cause ionization. There are now two electrons with low kinetic energy. This process is called impact ionization. This process happens repeatedly where the free electrons grow from the seed electrons exponentially. Eventually the material is broken down until a critical plasma density is reached and the dielectric material becomes absorbing. This is called the inverse Bremsstrahlung effect, where Bremsstrahlung is the emission of photons from excited electrons.

If the particle is constrained by the lattice's bonding energy, vibrations will be induced that spread throughout the structure; this is registered as heat. The incoming beam may interact with the evaporated material causing plasma formation and shielding effects. Thus, after a period of time sufficient energy is absorbed and transferred to the electrons, which then in turn heat the material which leads to melting and evaporation. Avalanche ionization can then occur for a time where collisions further ionize the material, releasing more material from the lattice.

Nanosecond ablation heats the specimen first to its melting point then to its vaporization temperature. The ablation depth per pulse is:

$$Z_a \approx \sqrt{at} Ln\left(\frac{F_a}{F_{th}}\right) \qquad [6.8]$$

where Z_a is the ablation depth, F_a is the absorbed fluence, F_{th} is the threshold fluence and *(at)* is the thermal diffusion depth. For $(at)^{1/2} = 0.5$ μm, a 20 ns pulse gives a typical threshold fluence of 4 J/cm^2. Nanosecond pulses (10^{-9}s) are considered to be a long *timescale*. The specimen is heated, then the main energy is lost due to heat conduction into the bulk of the material. Bulk heating may cause phase transformations to occur that may be harmful to the functioning of the material after machining.

As laser power intensity increases, the interaction between substrate and laser progresses through a number of characteristic phases with subsequent changes in machining characteristics. As the beam intensity is increased, the material begins to evaporate or ablate. Ablation starts when the surface temperature of the material exceeds its evaporation temperature. If the molten material is not sufficiently viscous to impede the ejection process, then increases in laser intensity lead to material removal by melt ejection and vaporization. Rapid heating of the substrate melts, vaporizes and then partly ionizes the vapor. The partly ionized vapor leaves the surface of the substrate at velocities up to 10^5 m/s. The net result of the production of a shockwave pushing the vapor and the melt droplets away from the substrate toward the laser beam is the creation of a recoil shockwave traveling away from the laser beam into the substrate. Several mechanisms will limit ablation rate. These include absorption and scattering of the laser beam by vapor and melt droplets being ejected during laser-beam interaction. The properties of the material and laser parameters determine whether evaporation, melt ejection or vaporization dominates the ablation process.

At higher intensities, in the order of 10^9 W/cm^2 and with pulse lengths of a nanosecond duration or shorter, the material is heated instantaneously beyond its vaporization temperature. Vaporization occurs within a very short time period at the start of the laser pulse. Energy dissipation of the laser pulse into the bulk of the material through thermal diffusion is slow relative to the pulse length. Before the surface vaporizes, the underlying material reaches its vaporization temperature. The temperature and pressure of the underlying material increases enormously, leading to the explosion of the workpiece material at the surface.

Recoil pressures above the surface of the material can be up to 10^5 MPa, leading to violent material ejection from the surface. High vaporization and plasma pressures above the surface increases the mechanical load placed on the material owing to a recoil shock-wave traveling away from the laser beam. Plasma-generated shockwaves can also be responsible for material ejection. However, during an ablative interaction, plasma is initiated at the surface. Plasma temperatures are in excess of 10^4 K and a plasma-material interaction, which is long in comparison to the short laser pulse, is established. This process is responsible for secondary heating of the substrate surface due to inverse Bremsstrahlung or photoionization. The interaction of plasma with ejected molten material is responsible for the deposition of molten material at the kerf. This always leads to the formation of a re-cast layer at the side of a laser-machined trench. The vaporization of this material, when ejected from the trench, is of paramount importance in order to maintain the precision required by micromanufacturing industries. The design of the nozzles has a profound effect on the way molten material is ejected from the machined trench and, consequently, on the way the ejected molten material is deposited and subsequently vaporized.

6.3.2. *Shielding gas*

Plasma is produced by laser ablation, which can shield the substrate, while the plasma itself can then be ionized. Plasma shielding can also be beneficial. The newly exposed material is protected by the plasma from interactions with the environment, e.g. oxidization. If the plasma is ionized and the conditions are appropriate, the plasma can acquire energy from the incoming beam such that it moves away from the surface and heads toward the source of the beam. This can reduce the energy coupling into the surface. In some cases, it blocks the path of the laser beam to the substrate and machining is halted. Various environmental gases can be used to influence this process, with the plasma being replaced with a shielding gas that requires a delivery system.

If the plasma is replaced with a gas that has a high ionization potential then ionization is made more difficult. In this way it is possible to have some control over this phenomenon. Often helium, argon, neon and oxygen gases are used; the type of gas used depends upon the reactions that take place. Much effort has gone into the design of nozzles that have different effects on the assist gas and on the machining of materials [JAC 02] [JAC 03]. Directional features include blowing a simple gas jet that blows the re-cast layer in a particular direction, interference with the beam, which, depending on the purity of the gas source, may cause the beam to reflect and become diffracted by gas contaminants, or successful removal of plasma.

6.3.3. *Nozzle designs for laser micromachining*

Many industrial processes that use gas jets to deliver fluid pressure to a workpiece surface employ simple conical nozzles. These nozzles are characterized by a minimum cross-sectional area at the exit of the nozzle. Inlet pressures greater than 1.89 atmospheres for diatomic gases produce under-expanded supersonic jets that exit from the nozzle. The axial pressure distributions delivered by these jets contain significant variations in magnitude within the first few nozzle diameters along the axis of the jet. Consequently, two restrictions are imposed by the choice of such a nozzle for laser micromachining applications. First, the exit of the nozzle must be very close to the workpiece in order to utilize the maximum impact, or shear force, available in the jet. Second, the complex nature of the axial pressure distribution demands that the stand-off distance between the nozzle exit and workpiece is very small in order to maintain uniform pressure at the workpiece surface. These drawbacks place limitations on the effective shear stress distribution, and hence the removal of molten material from the front of the kerf. Conventional high-speed nozzles have been used but they are associated with high shielding gas consumption and tend to generate high surface roughness features.

The alternatives to a conical nozzle are shown in Figure 6.2. The alternative designs include the use of a positive converging-diverging nozzle (de Laval) and a minimum length nozzle (MLN), which can create uniform supersonic jets with specifically designed flow fields with specific properties. One such property is a highly uniform and stable pressure that extends many nozzle diameters downstream along the centerline of the jet. In order to expand an internal steady flow through a duct from a subsonic to a supersonic regime the duct has to be convergent-divergent in shape. Both nozzles exhibit uniform free jet structures, especially de Laval, which, if correctly designed, produces a uniformly expanded jet that is free from shockwaves. Unfortunately, in laser material processing, small-scale de Laval nozzles with a very high length-to-diameter ratio may cause problems with optical delivery. This is especially a problem for micro-laser machining applications, where the desire for a minimum laser spot size means that short focal length optics is required.

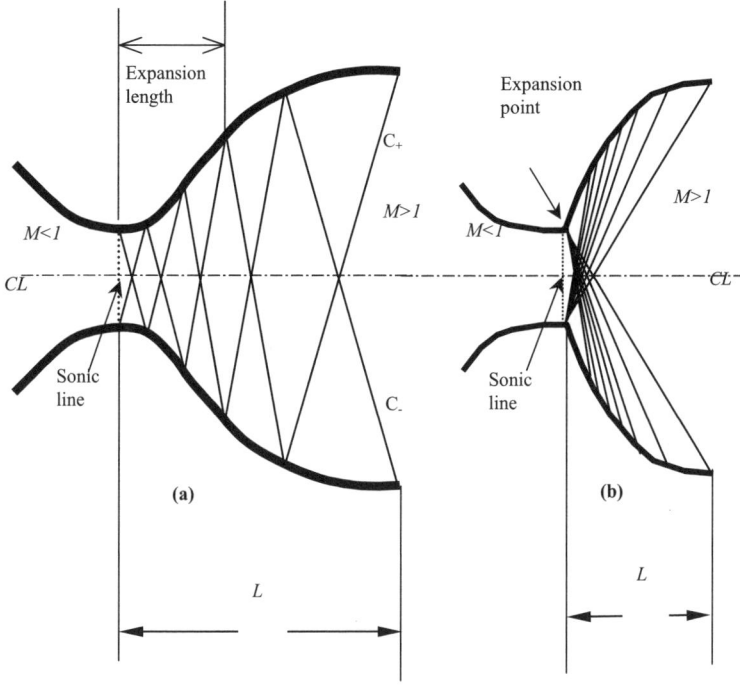

Figure 6.2. *Schematic diagram of supersonic nozzles: (a) de Laval; and (b) minimum length nozzle (MLN)*

Supersonic jets present significant drawbacks such as high mass flow rates and noise levels that are associated with jets of moderate size. It has been shown that sound pressure levels up to 150 dB at a radius of 30 nozzle diameters for 8 mm diameter nozzles are possible. However, miniaturized nozzles described in this study have avoided these problems by virtue of their small scale. Nozzles that have exit diameters of 1 mm produce characteristic supersonic jets with low mass flow rate and low noise levels. The majority of design criteria used in micromachining can be met by using an MLN, which were developed for use as a rocket nozzle, where minimum mass flow rates are a desirable property. This is also the case where rapid expansion is desirable, such as non-equilibrium flow in gas-dynamic lasers. Another advantage associated with minimum length nozzles is that boundary layer growth can be kept to an absolute minimum, in contrast to de Laval nozzles. This effect is noticeable when nozzle dimensions are very small. A comprehensive survey of existing literature concerned with minimum length nozzles has shown that axisymmetric and curved, sonic line MLNs have been limited to gas-dynamic laser applications and rocket engines. This type of nozzle eliminates the use of exotic gas mixtures at the laser-material interaction zone.

6.3.4. *Stages of surface melting*

There are a number of stages associated with surface melting. First, a melt front wave forms and begins to penetrate the surface. The pulse then finishes and no more energy can be put into the process. Then the advancement of the wave front ceases, indeed it retracts and begins to cool the material. Finally, re-solidification occurs; the volume that was formerly melted has now solidified with the new microstructure. However, there are problems, such as different temperatures in and around the mix that cause differences in surface tension, which promotes mixing forces to take control within the melt. These are called Marangoni forces and produce a texture on cooling the surface that may be undesirable, having caused residual stresses and cracking to occur. The advantages of laser surface melting means that bulk properties are unaffected, and in theory only a small surface depth is affected.

6.3.5. *Effects of nanosecond pulsed microfabrication*

If the shielding gas is oxygen, then a re-cast layer is usually produced when using nanosecond pulsed lasers. A thin oxide layer also accompanies the re-melt layer, which requires a secondary operation to remove it. The process of using assist gases is described in detail by Jackson *et al.* [JAC 02] [JAC 03], as are the reasons why it is so pronounced. Argon and air also tend to produce a significant re-cast layer when used in a controlled atmosphere. Experiments were undertaken to find out what effects the assist gas had on ablation or etch rates. It was found that adding an assist gas reduced the etch rate [JAC 02] [JAC 03]. In addition, the assist gas pressure greatly influences microfabrication using nanosecond pulsed lasers. Higher gas pressures were also found to produce a low etch rate, mainly because at high gas pressure the molten material was prevented from being expelled from the surface of the material.

The effects of varying inlet gas pressure on machining rates are shown in Figure 6.3, which shows the average etch depth per pulse achieved, when drilling a 1 mm thickness M2 tool steel plate. It is evident that the etch rate achieved was considerably higher than for no process assisted gas, which relies solely upon the vaporization of molten material.

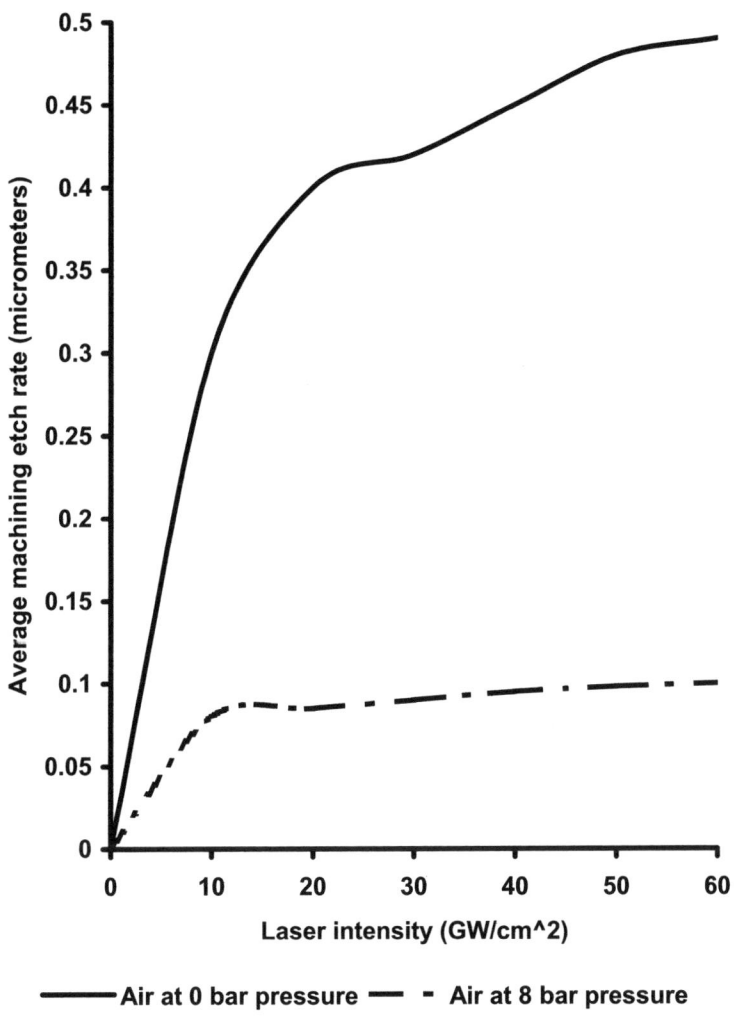

Figure 6.3. *Variation of the machining etch rate as a function of gas pressure*

High-speed photography of plasma initiation was undertaken using an Imacon 468 high-speed camera. The complete plasma life cycle was measured to be approximately 850 ns with no shielding gas. A significant change in the plasma geometry is evident when changing the gas jet pressure. In particular, distinct regions of distortion can be seen at 4 and 8 bar pressure. The distortions manifest themselves as a diamond shape at the top of the plasma and a longitudinal expansion at the base. These regions correspond to the positions achieved by the primary (normal) and the secondary (oblique) shockwaves. Immediately after a shock, the molecular density of the jet increases significantly, thus reducing the mean free path of molecules due to compression and hence increases ionization of the gases. This complex plasma structure will severely disturb beam propagation and affect the etch rates. At low shielding gas pressures, there is a pronounced removal of material that results in the formation of a trench in the workpiece material. The micrograph shows a typical cross-section of a laser-machined material with a correspondent re-cast layer. The re-cast layer is caused by suppressing the release of the molten material into the jet stream where it would subsequently be vaporized. This leads to the molten material being deposited at the sides of the machined trench. On further increasing the shielding gas pressure, the formation of oblique and normal shockwaves prevents the molten material from being removed from the trench and leads to the non-formation of a machined trench. As a result of a reduction in the machining etch rate, the amount of material re-deposited at the sides of the depression in the center is very small compared to the low shielding gas pressure regime. In all cases, the aspect ratio (L_H/D_H) is 9.32.

Low aspect ratios tend to produce the greatest amount of re-cast layers owing to the high shear stresses generated in the molten material, while high aspect ratios tend to produce depressed re-cast layers owing to the lack of shear stresses in the molten material. The increase in surface roughness as a function of the nozzle length-to-diameter ratio appears to show the advantages of using a high aspect ratio nozzle with a low-pressure inlet shielding gas jet. However, when the surface roughness of laser-processed samples was performed, we noticed that a range of surface roughness values existed for a variety of aspect ratios and inlet gas pressures. The range of surface roughness as a function of aspect ratio is shown in Figure 6.4, and as a function of inlet gas pressure in Figure 6.5. The variation in surface roughness is not easily explained. One explanation may be that the influence of interacting shockwaves prevents molten material from entering the jet stream above the interaction zone. Here, the laser beam normally makes contact with the jet of molten material in order to vaporize it. However, this is not the case when shockwaves in the gas jet interact with each other. This explains why, in some cases, the re-cast layer is larger than in other cases. Another explanation may be attributed over a longer period of time. As the laser beam establishes contact with the workpiece material, a dynamic plasma is initiated that causes interference with the action of the laser beam, thereby allowing the laser beam to deflect about a point where expulsion of the molten material is prevented. This means that molten material is not vaporized because the laser beam has been deflected away from the interaction zone. The material then solidifies at the side of the machined trench, which exaggerates the size of the re-cast layer. In both cases, time-dependent behavior of the laser interaction process appears to have produced a variation in the size of the re-cast layer. This has tremendous implications on the accuracy of micro-components manufactured using laser-based processing techniques.

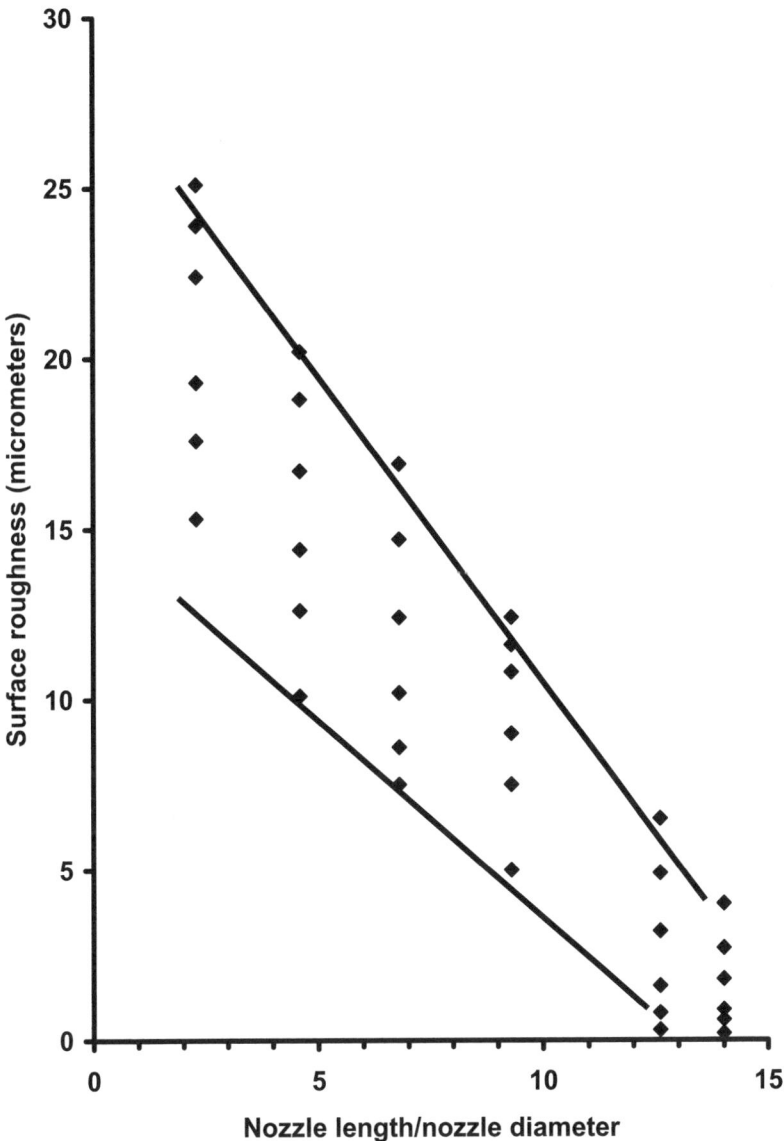

Figure 6.4. *Effect of using different nozzle length-to-diameter ratios on the surface roughness of machined trenches. The inlet gas pressure was 8 bar*

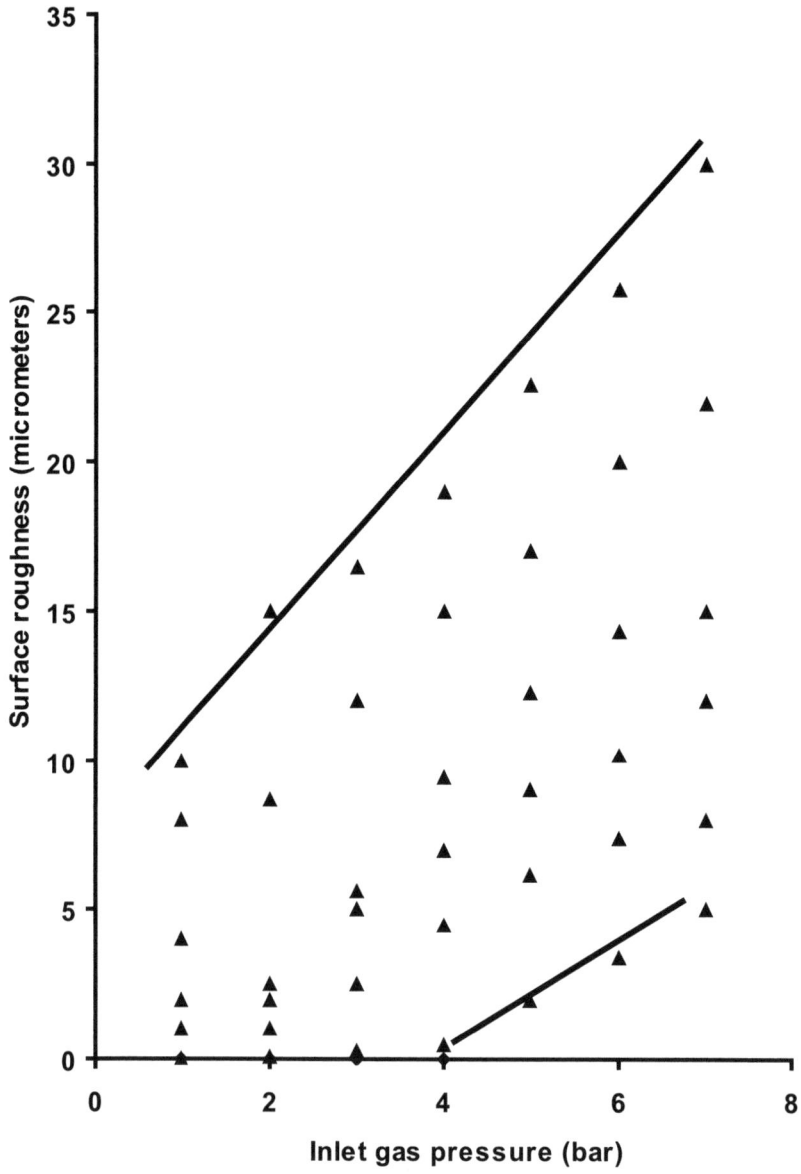

Figure 6.5. *Effect of using different inlet gas pressures on the surface roughness of machined trenches. The nozzle length-to-diameter ratio was approximately equal to 9*

6.3.6. Picosecond pulse microfabrication

In picosecond laser ablation, pulses are of the same *timescale* as it takes to transfer energy from electrons to the lattice of the material being machined. There is little heat conduction but a great deal of heat flow caused by free electrons. At the surface there is a solid-to-plasma phase; however, there is a liquid phase "inside" the material. The ablation depth per pulse is:

$$Z_a \approx \alpha^{-1} Ln\left(\frac{F_a}{F_{th}}\right) \qquad [6.9]$$

where α is the absorption depth. This liquid phase reduces the precision of the machining process. It has been shown that for processing metals, the pulse duration should be less than the electron-phonon thermalization time. For engineering materials like steel, copper, aluminum and iron, the thermalization time is around 10 ps. With shorter femtosecond pulses, deformation of the high intensity beam such as astigmatism coupled with non-linear interactions with the ambient gas cause unwanted effects. Diode pumped solid-state picosecond lasers are thus suitable for processing laser materials because they operate at high average power levels and repetition rates at around 100 kHz. An incident laser beam is absorbed by collisions with energetic electrons (electron collision time is around 100 fs) and free electrons, which results in heating of the lattice.

The electron-phonon relaxation time is greater than the electron collision time. In the case when the time is shorter than the electron-phonon relaxation time, the material dependent relaxation time is the dominant mechanism governing evaporation and solidification. For example the electron-phonon relaxation time for aluminum is around 100 ps. Ideally, to minimize unwanted thermal effects, the pulse duration should be less than the electron-phonon relaxation time. Therefore, the electron-phonon relaxation time of the material determines the extent to which thermal damage will be incurred. Therefore, picosecond laser micromachining is suited to cutting lines or patterns, drilling holes, surface structuring or milling. A comparison can be made with femtosecond lasers for the production of a mask. A 100 μm thick steel sheet 20 x 20 mm^2 takes 3,200 s to machine with a femtosecond system operating at 100 fs, 100 nm, 10 kHz and 0.37 W (37 μJ/pulse). The same mask can be produced to the same specification with a picosecond system operating at 1,064 nm, 50 kHz, 0.84 W (17 μJ/pulse) but the machining time in this case is only 570 s. If the repetition rate is doubled to 100 kHz the processing time can be further reduced by a factor of 12 with little sacrifice in quality when measured against parameters such as dimensional accuracy, minimal debris, minimal burr formation and a minimum HAZ.

144   Nano and Micromachining

Surface structuring or surface modification is possible with picosecond systems. Figure 6.6 shows an example where 25 μm grooves have been placed in 50 μm thick stainless steel. No burr formation or re-cast layer is observed, the power was 0.4 W (of 10 W) and the removal rate was 0.066 mm^3/min. Another application for surface structuring is the removal of thin layers (10 μm or less) that are structural features. Examples include direct delamination of metal layers, removal of thin indium tin oxide layers and laser honing. Typical powers are around 0.5 to 1 W and material removal rates of 10^{-3} mm^3/s are commonplace.

Lasers can be used to drill holes but special care must be taken. It is possible that because of reflection and diffraction effects, the hole may not have the same exit and entry diameter, nor will its walls be straight or parallel. To counter this effect, techniques such as helical drilling and polarization control with trepanning optics must be employed (Figure 6.6).

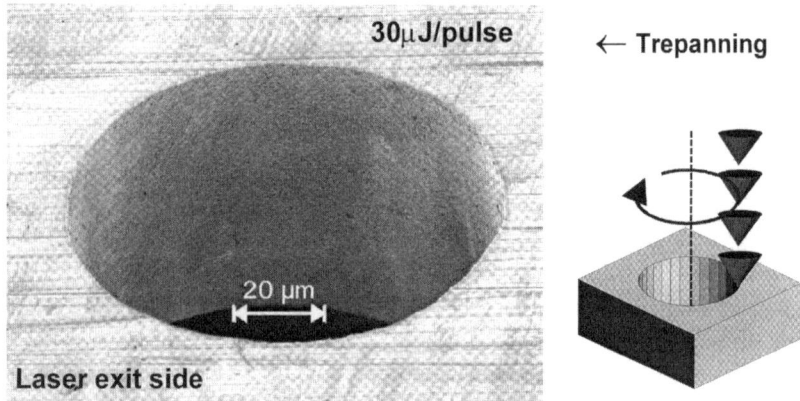

Figure 6.6. *Microdrilling of holes using a trepanning technique*
(courtesy of Lumera Laser)

Small holes around 50 μm in diameter can be produced with good edge quality, while no burr or debris is observed at a power of 2 W and a rate of 3 holes/s. Another example of laser-drilled holes is where an array of 30 μm holes are drilled in 25 μm thick steel foil at a rate of 2 holes/s at a power of 0.5 W (Figure 6.7). The holes have good edge quality, good circularity, were highly reproducible, and had minimal burr or debris formation.

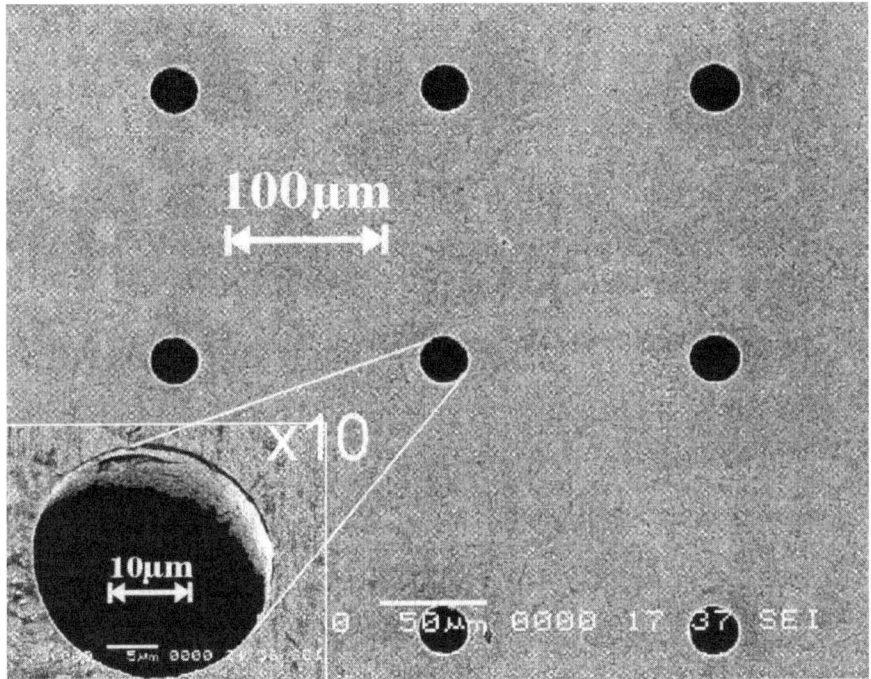

Figure 6.7. *Picosecond pulsed laser machined holes in stainless steel (courtesy of Lumera Laser)*

Where milling is concerned, the production of smooth μm surfaces can be achieved: for aluminum this rate is 1×10^{-3} mm^3/s; for steel it is 20×10^{-3} mm^3/s. An important application for lasers is the cutting of silicon wafers. The kerfs between chips must be minimal to ensure that the maximum density of chips are produced from the wafer. Machining is performed at 1.2 W with a removal rate of 3×10^{-3} mm/s and the requirements are good edge quality, minimal debris, minimal recast, and no crack formation. Drilling holes in silicon is important for high frequency technologies where electronic components pass through holes in silicon wafers. These holes cannot be produced in a satisfactory way by etching alone. Picosecond lasers with a beam diameter of 0.5-2 mm, using a power of 5 W and a time of 10 s are used to perform this function.

Ultrashort ps pulses have Terawatt power densities per square centimeter. Usually, free electrons are involved in absorption of laser energy. However, some materials do not have free electrons, but the energy of these ultrashort pulses is deposited by multiphoton absorption and electron impact absorption, or interband transitions. Hence, materials that are difficult to machine with a conventional laser

146 Nano and Micromachining

such as very hard or transparent materials like diamond or PTFE can be machined by ultrashort picosecond pulses. For example, PTFE has excellent properties for medical and electronic applications. Hence, microparts for these applications, which are difficult to mechanically machine, can now be machined using lasers. Processing of PTFE can be carried out at 10 W with a material removal rate of 0.2 mm^3/s. Laser machining of ceramics has also been shown to be successful. Ultrashort ps pulses produce good hole geometry due to non-thermal ablation with no thermal cracks present. This can be carried out at 10 W at 8 holes/min. Similarly, holes can be drilled in borosilicate glass. The holes have a 1 mm diameter, are 140 μm thick and the edge quality is excellent with no cracking and was carried out at 2 W at 20 holes/min. Table 6.2 shows typical material removal rates for picosecond pulsed lasers machining a wide variety of engineering materials.

6.3.7. *Femtosecond pulse microfabrication*

When machining with femtosecond pulsed lasers, an electromagnetic wave interacts with the particle, multi-photon absorption occurs, and there is no liquid phase. In theory the pulse and the evaporation interaction occur before the next pulse impacts the material. Bounded electrons of the material can be directly ionized by multi-photon absorption. Equally, photons with energy, $h\nu$, can be absorbed to ionize the atom, although for this to happen the ionization potential or band gap must be surmounted. The electron does not have to wait for energy to be supplied to it. The pulse is so short that the energy is supplied at a rate where direct ionization occurs. Long pulses have low breakdown field strength whereas short pulses have high breakdown field strengths. The breakdown fluence threshold can be defined as the fluence at which there is a 50% probability that a laser of this pulse width and fluence will cause breakdown to occur. The field strength is determined by the density of the field lines of the electromagnetic wave. For long pulses the field lines are spread out and for short pulses the field lines are closer together. For long pulses, avalanche breakdown thresholds trigger the material removal process but are random in nature; for short pulses, multiphoton absorption is the dominant breakdown trigger and occurs rapidly. For long pulses, statistical variations occur as a result of the distribution of seed electrons required for the onset of ionization, which can take some time for avalanche ionization to be triggered.

Material	Material removal rate (mm^3/s)
Aluminum	0.001
Steel	0.02
Copper	0.001
Titanium	0.06
Silicon	0.005
Ceramic	0.08
Glass	0.14
WC	0.03
SiC	0.009
PTFE	0.2

Table 6.2. *Material removal rates for materials using picosecond pulsed lasers (courtesy of Lumera Laser)*

For multiphoton absorption there is no randomness as there is no reliance on seed electrons to initiate the process. Any electron, free or otherwise, can be ionized by multiphoton absorption. This is why femtosecond lasers are capable of machining any material. The ablation process is completed by avalanche ionization even if it is triggered by multiphoton absorption. Once the plasma density reaches a critical level, ablation can occur. Electrons absorb energy by collisions and are heated to extremely high temperatures. Simultaneously, electrons transfer energy to the ions and the lattice, which tends to heat the surface. Energy transfer depends on pulse duration and the coupling coefficient. Short pulses avoid a long wait before the critical level is reached because they do not rely on free electrons while waiting for the critical level to be reached are avoided. For long pulses the energy transfer from electrons to ions is efficient. The heat diffusion zone is larger and as the pulsewidth decreases the field strength increases and electrons reach a much higher temperature than the ions or the lattice. Vaporization occurs rapidly and the HAZ is very small.

An interesting observation is that because multiphoton absorption does not depend on the presence of free electrons any material can be processed. Materials that are transparent to the wavelength, such as glass, can be machined.

In femtosecond ablation, no energy is transferred to the lattice, i.e. all energy is stored in a thin surface layer. The ablation depth per pulse is given by:

$$Z_a \approx \alpha^{-1} Ln\left(\frac{F_a}{F_{th}}\right) \qquad [6.10]$$

Note that equation [6.10] shows that there is no ablation observed at the threshold fluence. For α^{-1} = 10 nm, the threshold is typically 0.1 J/cm². If the energy stored in the thin surface layer is greater than the specific evaporation temperature then vigorous evaporation will occur after the pulse. Ideally, the ablation fluence should be around three times greater than the threshold to remove an irradiated layer, thickness α^{-1}. Ablation occurs from the solid to the plasma phase. The plasma expands rapidly and is expelled from the surface, and because there is no time for heat transfer to occur, it is therefore a very precise ablation process. This implies that re-cast layers should no longer form on the surface of the material being machined.

Figure 6.8 shows a comparison between using a nanosecond pulsed laser and a femtosecond pulsed laser to micromachine. It is shown that the left-hand figure has a pronounced re-cast layer, which is produced using a nanosecond pulsed laser. The right-hand micrograph has no re-cast layer, indicating that it has been machined using a femtosecond pulsed laser.

Nanosecond machining often produces re-cast layers, resulting in a loss of dimensional accuracy. There is also a HAZ and post-processing is required to generate the final micropart. Femtosecond machining results in no re-cast layers being formed and clean holes and tracks, which can be produced without secondary processing. Femtosecond laser machining can produce micro-fluidic channels in a variety of materials. Figure 6.9 shows the machining of trenches in the borosilicate glass that is used for microfluidic applications.

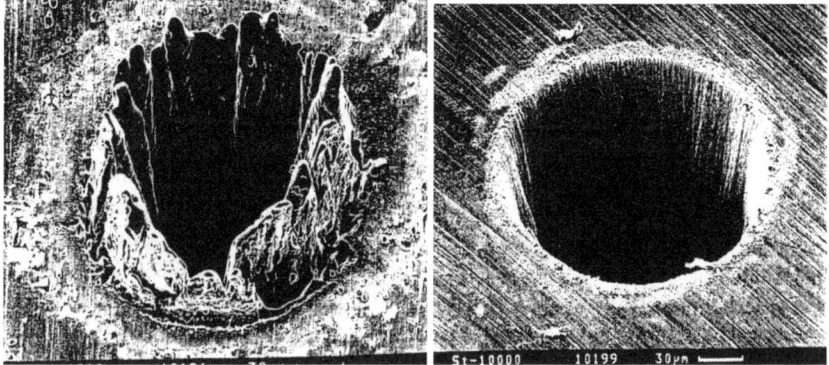

Figure 6.8. *Effects of using nanosecond and femtosecond pulsed lasers to machine an engineering material. The material in the left image is machined using a ns pulsed laser and the right image is machined using a fs laser. Note the lack of a recast layer using fs lasers [CHI 96]*

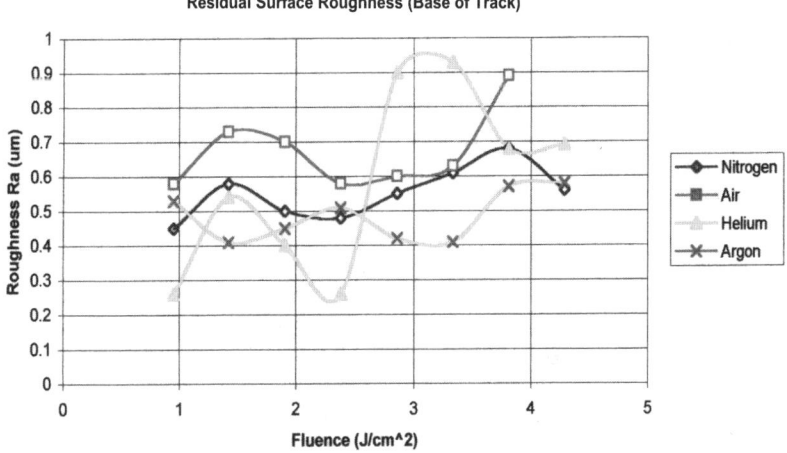

Figure 6.9. *Effect of assist gas on the surface roughness of aluminum using a femtosecond pulsed laser*

The effects of using different assist gases were investigated. As is the case for nanosecond pulsed laser ablation, the ablation rate is affected by the type of assist gas used at various levels of fluence. Figure 6.9 show the effects of using assist gases on the ablation rate and on surface roughness.

Note that the roughness of the machined trenches is identical to the roughness of the original material. This is because the femtosecond laser removes a very small amount of material. Thus, the same vertical distance is removed and an identical profile to that at the surface is created at the base and side-walls of the trench. Air produces relatively good results with small amounts of material build-up around the edges. Argon produces a powdery residue around the track, while nitrogen yields poor results, causing a build-up of material around the machined trench or channel. Helium produced the cleanest results. All other gases produced a powder that was black, helium on the other hand, produced a silvery deposit. The black powder produced was thought to be a nanoscale powder with deposits so small that they are below the wavelength of light and therefore no interaction is possible, resulting in the black appearance. Clearly the assist gas has some effect but this has not yet been quantified. Processing with femtosecond lasers may provide an economical way to produce nanometal powders.

6.3.8. Effects of femtosecond laser machining

The following results were achieved in an ambient airflow generated at 150 µJ/pulse, with a spot diameter of approximately 30 µm and slightly elliptical in shape. The overlap between lines is 10 µm. The textured surface produced at the base of the cube increases in variability with increasing number of scans. It can be seen that the relief of the sidewalls becomes more severe as the number of over-scans increases; this may be attributed to the nature and shape of the Gaussian profile of the beam.

The textured surface appears to be globules of metal left behind by the ablation of the surrounding matrix. Alternatively, the structured surface could be condensed droplets of re-cast that have fallen and solidified. Alternatively, the texture could be localized solidified melt pools created by the heat from the plasma (heat from the beam directly evaporates material; therefore, if present, secondary heating effects are likely to arise from the plasma). It could be reasoned that the best quality machined surface is machined with a combination of a fast scan speed to reduce texturing and many over-scans to achieve a reasonable depth. Further experiments are required to conduct these experiments at varying power levels, where it could be reasoned that at very low power levels the formation of texture is reduced. The inherent problem with laser micromachining is the shape and nature of the Gaussian beam profile. This problem is manifested in trenches where the walls of the machined slot are not parallel.

6.4. Laser nanofabrication

Laser nanofabrication is being used to enhance processes such as atomic force microscopy and molecular beam epitaxy by dispersing accumulated atoms that form "islands" of atoms from randomly deposited sources. Using the forces exerted by laser light tuned to near-atomic resolution, an array of atomic lenses is formed that concentrates atoms in an array of lines with lengths of around 30 nm. This novel form of nanofabrication can create nanostructures without the use of a resist or other pattern transfer technique. The technique has been used to focus sodium, aluminum and chromium atoms by the force created by the nodes of the standing light wave that touches them. Laser-focused atomic deposition has been used to create nanostructures of chromium atoms that are 60 nm in height and 28 nm in width. The basic principle of laser focused atomic deposition [McC 93] is that atoms pass through a near-resonant laser standing wave as they deposit to the surface. An included dipole moment on the atom interacts with the laser light to cause a force to act toward the nodes of the standing wave. The resulting nanostructures are as small as 28 nm and are spaced at half a wavelength [GUP 95].

The deposited features can also be used as pattern masters, which have been successfully used for polymer molding processes that are suitable for developing into a nanomanufacturing process. In other approaches to solve nanofabrication problems, metastable rare-gas atoms have been shown to be an effective exposure tool for lithographic processes. Metastable atoms can be focused using a laser in ways similar to chromium atoms and be manipulated to produce nanofeatures. Figure 6.10 shows an array of AFM images depicting structures produced using laser-focused lithography [MES 03].

When coupled with reactive ion etching procedures, laser based nanofabrication of chromium nanostructures can be deposited to create trenches and channels that may prove useful for nanofluidic applications. Figure 6.11 shows the results of combining the two methods. Another interesting development in the use of lasers to fabricate products at the nanoscale was developed by Grier at the University of Chicago. Grier used a laser as an optical tweezer to move material from one place to another that was originally in the form of a suspension, or colloid. The beam of light is split up into an array of beams so that particles become trapped inbetween the beams of light. Arrays of nanostructures can be manipulated in this way, which can lead to a new form of nanomanufacturing when coupled with established processes. Laser nanofabrication is at a very early stage of development at the moment and will probably need to be combined with other processes to form hybrid nanofabrication and nanomanufacturing processes [STI 03]. A large number of researchers are currently investigating the development of hybrid forms of laser nanofabrication.

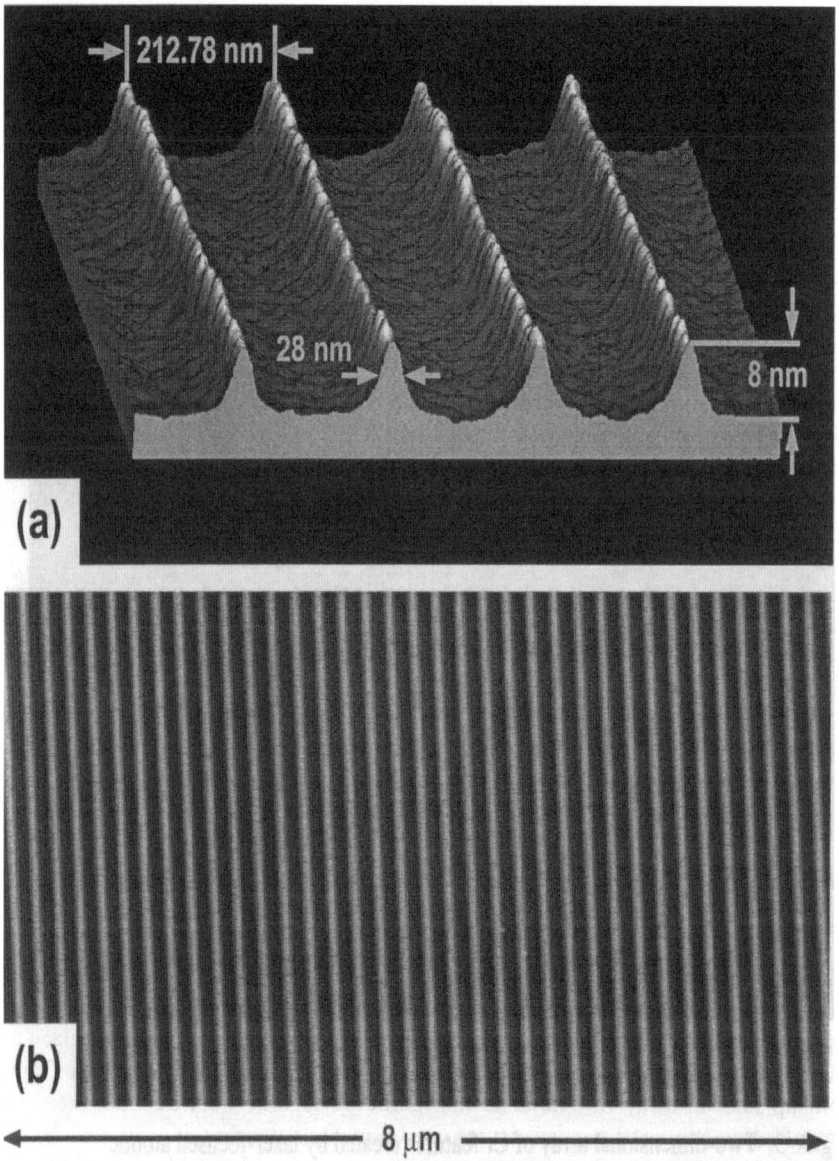

Figure 6.10. *Atomic force microscope images of laser-focused chromium nanostructures: (a) 3D image of nanoscale features deposited to silicon; and (b) image of 60 nm high nanofeatures on sapphire*

Figure 6.11. *Nanostructures formed when combining reactive ion etching with laser-assisted deposition of chromium atoms: (a) 66 nm wires formed when chromium contrast is highest; (b) uniform trenches in a silicon substrate formed at medium contrast; and (c) narrow trenches formed when contrast is low*

6.5. Conclusions

The use of lasers to fabricate products at the microscale is well established. CFD studies of supersonic jets interacting with holes of varying aspect ratio have highlighted complex shockwave structures present during laser micromachining, which enhances the shear stress distribution in the molten pool, something required to physically remove molten material and send it into the jet stream above the surface of the material. Experimental observation has revealed reduced etch rates at high gas pressures, which is coincident with dense plasma formation within the shocked gas stream above the surface of the workpiece. A reduction of gas pressure to near atmospheric pressure reduced detrimental plasma effects and reduced the height of re-cast layers once dominant at high pressure. A simple design rule used for laser micro-machining is to reduce the length to diameter ratio of the nozzle. The nozzle is used with shielding gas pressures in the range 0 to 0.5 bar in order to prevent the formation of shockwaves in the dynamic plasma, a process responsible for the formation of re-cast layers. The development of processing at the microscale continues to be dominated by the application of femtosecond pulsed processing of engineering materials. However, the development of attosecond pulsed lasers may eliminate the problems created by the formation of a plasma that accompanies the more traditional lasers currently used for microfabrication. Nanofabrication processes that use lasers to create useful nanofeatures are still in their infancy of development. Although the rapid strides made in laser manipulation at the nanoscale may see the development of laser-based nanomanufacturing processes in the not too distant future.

6.6. References

[CHI 96] CHICHKOV, B.N., *et al.*, "Femtosecond, picosecond and nanosecond laser ablation of solid surfaces", *Applied Physics*, Vol. A63, 1996, p. 109-115.

[GUP 95] GUPTA, R., *et al.*, "Nanofabrication of a two-dimensional array using laser-focused atomic deposition", *Appl. Phys. Lett.*, Vol. 67, 1995, p.3718.

[JAC 02] JACKSON, M.J., *et al.*, "Evaluation of supersonic nozzle designs for laser micro-machining of high speed steels", *Transactions of SME – Journal of Manufacturing Processes*, Vol. 4, 2002, p. 42-51.

[JAC 03] JACKSON, M.J., *et al.*, "Micromachining of high chromium content steel under controlled gas atmospheres", *Transactions of SME – Journal of Manufacturing Processes*, Vol. 5, 2003, p. 106-117.

[JAC 03] JACKSON, M.J., *et al.*, "Laser micro-machining of chromium rich die steel under controlled atmospheres", *Proceedings of the Institution of Mechanical Engineers (London): Part B – Journal of Engineering Manufacture*, Vol. 217, 2003, p. 553-562.

[McC 93] McCLELLAND, J.J., *et al.*, "Laser-focused atomic deposition", *Science*, Vol. 262, 193, p.87.

[McG 02] McGEOUGH, J.A., *Micromachining of Engineering Materials*, Marcel Dekker, New York, 2002.

[MES 03] MESCHEDE, D and METCALF, H, "Atomic nanofabrication: atomic deposition and lithography using laser and magnetic forces", *J. Phys. D.: Appl. Phys.*, Vol. 36, 2003, R17-R38.

[STE 98] STEEN, W.M., *Laser Materials Processing*, Second Edition, Springer, New York, 1998.

[STI 03] STIX, G, "Hands of light", *Scientific American*, August 2003, p. 30-31.

Chapter 7

Evaluation of Subsurface Damage in Nano and Micromachining

7.1. Introduction

Many machining processes will induce surface/subsurface damage in the material. As shown in a ground silicon wafer in Figure 7.1, surface damage means that there is some damage on the surface of the material, like surface roughness; subsurface damage (SSD) means that there is a damaged layer below the surface, like subsurface cracks, dislocations, residual stress, etc. [BIS 94, DEG 08, HAD 85, QUI 01]. In order to avoid failures in the final device production, this SSD layer must be eliminated by the subsequent processes. Therefore, it is necessary to detect the subsurface damage depth caused by the machining processes and use this knowledge to optimize the subsequent processes.

There are many applicable techniques for characterizing SSD in nano and micromachining. In general, they can be classified into two categories: destructive and non-destructive. "Destructive" means that the samples will be destroyed during the measurement. However, in order to enhance the cost-effectiveness in manufacturing, it is desirable to have non-destructive evaluation (NDE) methods to measure SSD. This chapter investigates nano- and micro-scale subsurface damage evaluation technologies, with the emphasis of their applications in MEMS.

Chapter written by Jianmei ZHANG, Jiangang SUN and Zhijian PEI.

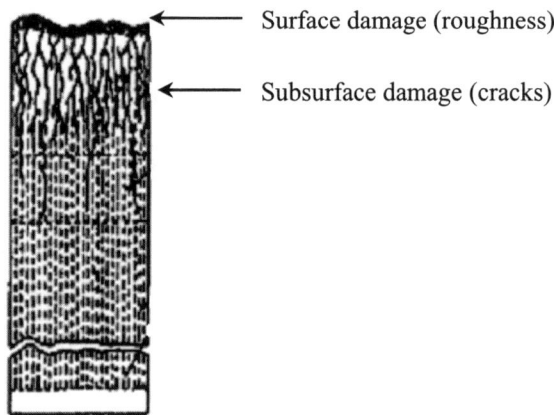

Figure 7.1. *Surface damage and subsurface damage layer in ground silicon wafers (after [HAD 85])*

7.2. Destructive evaluation technologies

7.2.1. *Cross-sectional microscopy*

For cross-sectional microscopy, the samples must be appropriately prepared, before observation under an optical microscope.

Normally, sample preparation consists of four general steps [PEI 99]: cleaving, sanding, polishing and etching. First, cleaving is done perpendicularly to the machined surface. Second, the surface of the cleaved sample is wet-sanded to remove enough material from the interested cross-section to ensure that any damage incurred during cleaving is removed. Third, the test surface is refined by polishing. The polished surface should also remain flat and perpendicular to the machined surface. Finally, the test surface is placed into "Yang" solution [YAN 84] (H_2O: HF 49%: CrO_3 = 500 ml: 500 ml: 75 g) for 5 seconds at room temperature, which will make the subsurface cracks more discernable for microscopy observation. Sample preparation is vital for characterizing the true subsurface cracks.

Cross-sectional microscopy cannot only assess the depth of subsurface damage, but can also provide information about subsurface crack configurations, which may help to explain the formation of different subsurface cracks. However, this technology also has some drawbacks. The main drawback is that it can only reveal the damage information for specific areas on the machined surface. If the deepest SSD is not among the tested areas, it will not be possible to capture it. Additionally, the sample preparation process is very tedious. Furthermore, the sample preparation process itself will generate cracks/damage if not performed properly.

7.2.2. Preferential etching

The procedure of preferential etching is simple. First, chemical mechanical polishing (CMP) is used to remove the specific layer from the surface. Second, an etching operation is carried out, making the defects visible by marked etch pits. Mchedlidze *et al.* used Schimmel-I solution [SCH 79] as etchant at room temperature for 5 minutes [MCH 95]. More often, the "Yang" solution is used as etchant to develop elongated rectangular, ellipsoid or circular etch pits in the surface. Third, the count of etch pits provides a quantity for damage intensity. Therefore, the SSD information can be obtained by using preferential etching technology [TON 90, TON 94].

Compared with cross-sectional microscopy, a major advantage of preferential etching is that the sample preparation is easier and less expensive. However, it does not give a quantitative analysis of the damage, and can hardly distinguish different types of defects, such as subsurface cracks and dislocations. Moreover, it can only obtain damage information for a specific layer. If the damage information of different layers is needed, step etching [STE 86] should be preferred.

7.2.3. Angle lapping/angle polishing

The procedures of angle lapping and angle polishing are almost the same, with the exception of the second step [JEO 00, TON 97]. First, the test samples are cleaved. Second, the samples are lapped (in angle lapping) or polished (in angle polishing) under an angle of about 5°. Third, in order to observe the damage clearly, oxidation is performed on the tilted surface. Fourth, selective etching is made on the tilted surface with "Yang" solution to make the defects visible. Finally, the etching pits are counted in the tilted surface of the samples to determine, via their distribution, the SSD depth. Figure 7.2 shows how the depth of SSD is assessed by angle polishing technology.

In the research of Jeong *et al.*, dry oxidation and wet oxidation were performed and the results were compared [JEO 00]. It was found that the damage was more notable for the samples processed through dry oxidation (at 800°C for 4 hours) than for those processed through wet oxidation (at 1,100°C for 80 minutes).

Angle lapping or angle polishing is also limited: it can only reveal the damage information of specific areas on the machined surface. Yet, the deepest SSD may not occur among the tested areas, and it will be missed. Furthermore, the sample preparation process involves many steps.

160   Nano and Micromachining

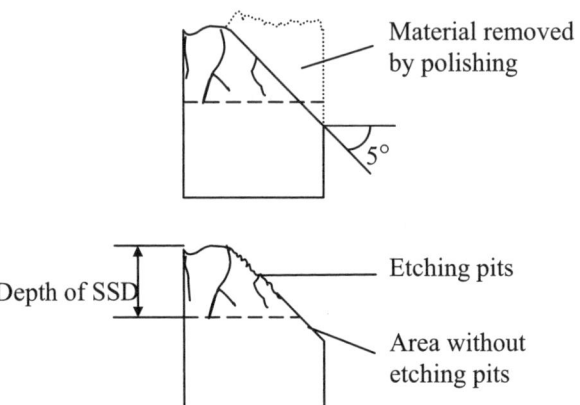

Figure 7.2. *Depth of SSD measured by angle polishing (after [TON 97])*

7.3. Non-destructive evaluation technologies

7.3.1. *X-ray diffraction*

7.3.1.1. *Introduction of X-ray diffraction*

X-ray diffraction is widely used as a non-destructive method of determining residual stress from the lattice deformation of a crystal. A crystal lattice is a regular 3D distribution (cubic, rhombic, etc.) of atoms in space. When X-rays come in at a particular angle, they are reflected specularly (mirror-like) from the different planes of crystal atoms. However, for a particular set of planes, the reflected waves interfere with each other. A reflected X-ray signal is only observed if Bragg's condition for constructive interference is met.

Figure 7.3 illustrates how X-ray diffraction works. Figure 7.3(a) shows X-rays incident upon a simple crystal structure, Bragg's condition is met for both ray A and ray B. Figure 7.3(b) shows the diffraction geometry. The extra distance traveled by ray B must be an exact multiple of the radiation wavelength. This means that the peaks of both waves are aligned with each other. Bragg's condition can be described by Bragg's Law: $2d\sin\theta = m\lambda$, where d is the distance between planes, θ is the angle between the plane and the incident (and reflected) X-rays, m is an integer called the order of diffraction and λ is the wavelength [HEC 02].

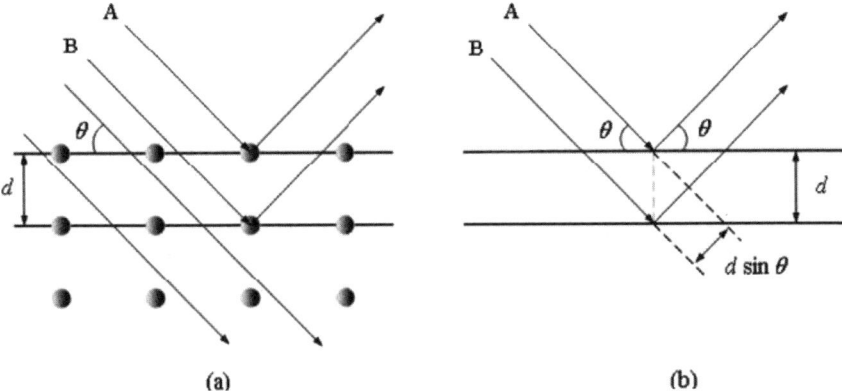

Figure 7.3. *Illustration of X-ray diffraction: (a) X-rays A and B incident upon a crystal; and (b) diffraction geometry (after [ONL 00])*

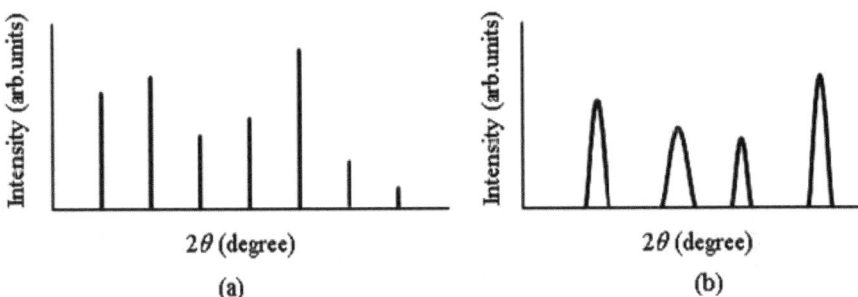

Figure 7.4. *Illustration of X-ray diffraction patterns: (a) for a perfect crystal; and (b) for an imperfect crystal (after [ONL 00])*

Figure 7.4(a) shows a standard X-ray diffraction pattern with aligned sharp peaks for a perfect crystal structure. During machining processes, residual stress may occur as a result of non-uniform, permanent 3D changes in the material. These changes usually occur as plastic deformation and may also be caused by cracking and local elastic expansion or contraction of the crystal lattice [BIS 94]. In these cases, the crystal structure is no longer perfect (d is changed) and the diffraction peaks are broadened and shifted, as shown in Figure 7.4(b). Thus, by examining the changes of the X-ray diffraction pattern, the residual stress can be characterized. Furthermore, the related defects can possibly be identified.

7.3.1.2. Application of X-ray diffraction

The main restriction on the applications of conventional X-ray diffraction is the low penetration depth of X-ray into the workpiece material. Because the strain can only be measured within the irradiated surface layer, with low penetration depth of X-ray, only the stress close to the surface can be detected quantitatively. Tonshoff *et al.* used a high resolution X-ray diffractometer (shown in Figure 7.5) to detect the damaged layer of machined silicon wafers [TON 90]. This diffractometer consists of a four-crystal monochromator that produces a highly parallel and monochromatic incident beam so that a high resolution of X-ray diffraction can be achieved. However, in general the spatial resolution of X-ray is much poorer than with spectroscopic techniques presented in section 7.3.2.

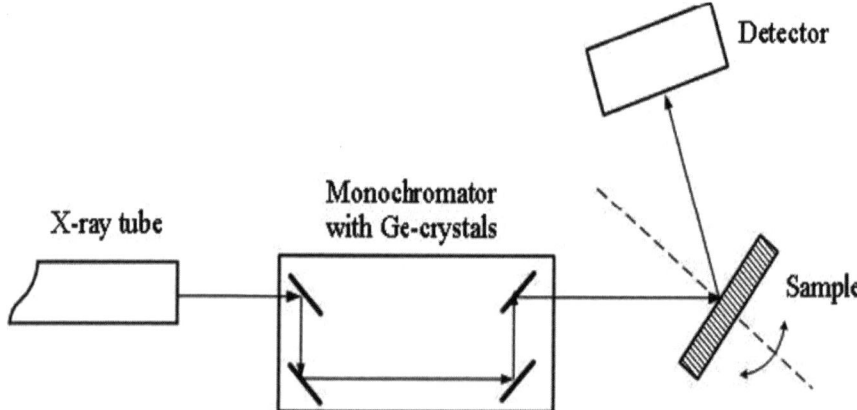

Figure 7.5. *Illustration of high resolution X-ray diffraction system (after [TON 90])*

The X-ray diffraction pattern of ground silicon wafers showed an increased diffraction intensity in "the wings of the Bragg-peak" caused by the damaged layer. Compared with that of the ground wafer, the diffraction pattern of a polished "perfect" wafer showed a considerably smaller width, that is, half the maximum of the Bragg-peak (shown in Figure 7.6). Bismayer *et al.* reported that the quantitative determination of residual stresses in silicon wafers could be achieved by the X-ray diffraction technique [BIS 94].

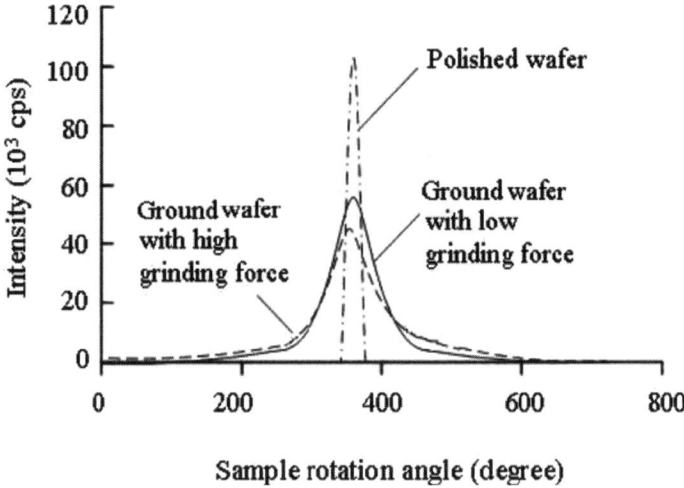

Figure 7.6. *X-ray diffraction pattern of silicon wafers (after [BIS 94])*

Mikulík *et al.* demonstrated the capability of high-resolution X-ray diffraction techniques with synchrotron radiation sources to inspect the structural perfection in semiconductor wafers [MIK 02]. They applied X-ray diffraction to visualize and characterize the defects (dislocations and microcracks) in semiconductor wafers (in particular, silicon wafers and GaAs wafers) induced by growing, cutting and grinding. The number and the sharpness of observed Pendellösung fringes depends on the crystal perfection. Figure 7.7 shows the SSD characterization of silicon wafers by Pendellösung fringes. Fringe visibility is the best for the "perfect" monitor wafer. No fringes are visible on the ground wafer. Etching-off or polishing-off (by CMP) only a surface layer of 5 μm in thickness after grinding is sufficient to remove the SSD significantly, resulting in a partial restoring of Pendellösung fringes.

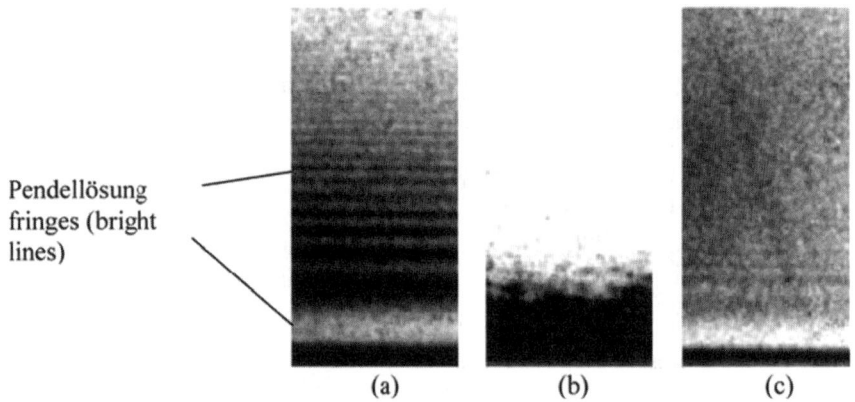

Figure 7.7. *SSD characterization by Pendellösung fringes:
(a) "perfect" monitor wafer (20 fringes); (b) ground wafer (no fringe);
and (c) ground and 46 μm etched wafer (7 fringes) (after [MIK 02])*

X-ray diffraction has been used to study silicon-on-insulator structures by Popov *et al.* [POP 01] and to study silicon carbide on insulator structures by Milita *et al.* [MIL02]. Their experimental results showed that X-ray diffraction could characterize the lattice defects in these wafers and quantify these lattice defects and related stresses at the bonding interface.

7.3.2. *Micro-Raman spectroscopy*

7.3.2.1. *Introduction of micro-Raman spectroscopy*

The principle of the Raman effect is based on an inelastic light-scattering process. When the light is scattered from a molecule, most light (photons) is elastically scattered, that is, most of the scattered light has the same frequency and wavelength as the incident light. However, a small fraction of light is scattered at optical frequencies different from and usually lower than the frequency of the incident light. This inelastic scattering of light is called the Raman effect. Further, the microscopic changes (long- and short-range disorders), impurities and residual strains strongly influence the Raman spectrum of a material. These microscopic changes lead to changes in photon frequencies, broadening of Raman peaks and breakdown of Raman selection rules. When a beam of monochromatic light passes through a crystal, the Raman effect occurs. Thus, the presence of SSD as well as residual stresses can be detected by this technology [YAN 04].

The basic experimental setup for a micro-Raman spectroscopy system is shown in Figure 7.8 [BIS 94].

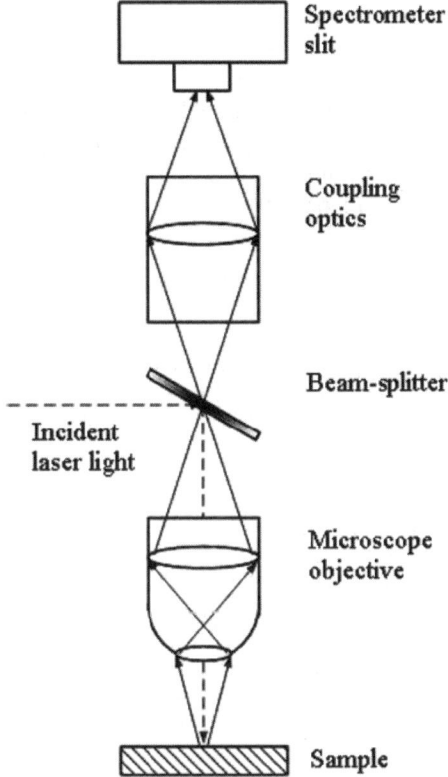

Figure 7.8. *Illustration of micro-Raman spectroscopy (after [BIS 94])*

The laser light is focused on the sample through a microscope (often a con-focal microscope where a spatial pinhole is used to eliminate the light, out of focus, thus increasing the contrast), resulting in a spot size down to about 1 μm (spot sizes as small as 0.3 μm can be obtained if an oil immersion objective is used). Generally, the spatial resolution of micro-Raman can be 1 μm [WOL 96]. The scattered light of the sample is collected using the same microscope focused on the entrance slit of a spectrometer (detection device). The sample can be mounted on a computer-controlled X–Y stage, which allows the sample to be scanned in small steps (typically 0.1 μm) in a given direction. The total collection volume depends on the light-scattering properties of the material being analyzed [HAR 04]. Furthermore, the light penetration depth changes with different wavelengths. Therefore, it is possible to probe different depths of the sample by varying the wavelength [SPA 88, SPA 93].

7.3.2.2. Application of micro-Raman spectroscopy

Micro-Raman spectroscopy is used for analysis of phase transformation and residual stress in machined silicon wafers. Sparks and Paesler used micro-Raman spectroscopy to measure residual stress in a shallow groove (50 µm) in silicon produced by single point plunge cuts [SPA 88]. Bismayer *et al.* used micro-Raman spectroscopy to measure the silicon wafers machined by ID sawing, grinding and polishing (the polished wafer was used as the reference) [BIS 94]. Compared with the polished wafer, different Raman peak positions and intensities were observed for the ground wafer (shown in Figure 7.9).

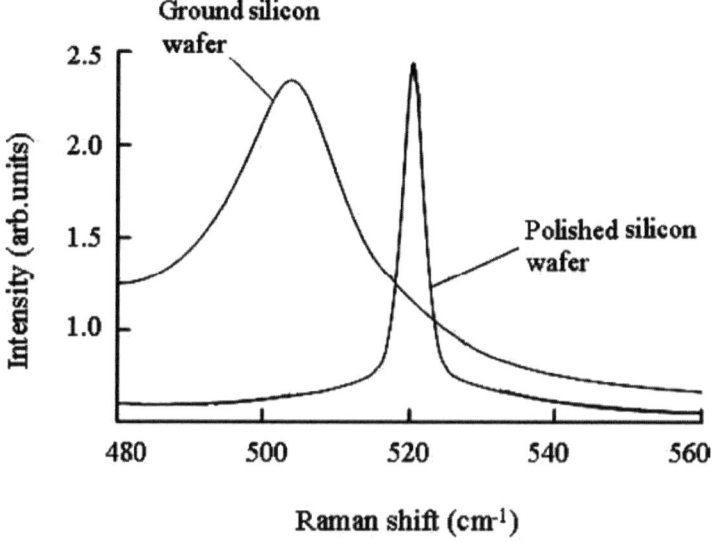

Figure 7.9. *Raman peaks of silicon wafers (after [BIS 94])*

This Raman peak shift was caused by the SSD induced by high stresses left in the wafer during the machining process. The damage depth could be obtained using depth-profiling techniques outlined by Sparks and Paesler [SPA 93]. The phase transformations in silicon within the penetration depth could also be analyzed using micro-Raman microscopy. The penetration depth changes with the laser beam wavelength, which allows data acquisition from depths of <0.1 µm to 10 µm [GOG 99].

The greatest advantages of micro-Raman spectroscopy are its high resolution, the simplicity of the experimental setup and the short time required for obtaining data

(data acquisition is automated and it takes 1 to 10 seconds to acquire a spectrum at an individual location).

7.3.3. *Laser scattering*

7.3.3.1. *Introduction of laser scattering*

With laser scattering technologies, when a laser beam directly illuminates a sample, the amount and distribution of back-scattered light could be monitored. The basic principle of light scattering technology is that different microstructures in a sample may cause a change in the scattering characteristic of the reflected light. Especially, a sudden change in the light intensity would occur when the incident light encounters a defect [WAN 00].

Figure 7.10 illustrates a laser scattering system. A laser beam passes through a pinhole and illuminates a small area of the sample. Light is reflected from the illuminated spot on the sample to the objective, where it is directed by a beam splitter toward the confocal pinhole aperture. The pinhole positioned in front of the detector gives the system its confocal property by rejecting light originating from neighboring focal planes. Light rays from an unfocused plane are blocked from reaching the detector. However, all light rays originating from the focal plane pass through the pinhole aperture and are collected by the detector. The ability to finely discriminate between the light rays originating at the focal plane and those originating at the unfocused planes makes a laser scattering system a powerful tool in generating depth-precise images [STO 90]. Usually, objective lenses with a high numerical aperture (NA) are used to provide good resolution in the x-, y- and z-directions.

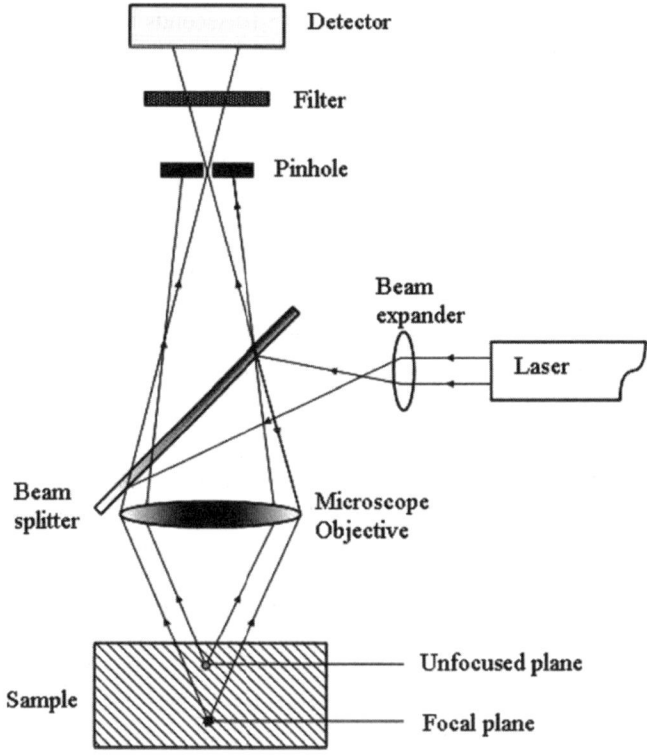

Figure 7.10. *Schematic diagram of a laser scattering system (after [STO 90])*

Laser scattering technologies cannot assess SSD on all materials. Only those having appropriate "skin depth" are suitable for laser scattering. When the light is incident on the surface of a material, it can penetrate beneath the surface for a distance known as "skin depth" because of the finite absorption of the material [STE 94, STO 90]. For metals, the "skin depth" is at most a few nanometers for visible wavelengths, thus scattering can be considered to emanate entirely from the surface. Therefore, laser scattering cannot be used to measure SSD in metals. For silicon-nitride ceramics, the "skin depth" was found to be from tens to hundreds of microns [SUN 99], so laser scattering can be applied to measure SSD within that depth range. For silicon wafers, Zhang *et al.* found that silicon wafers have the appropriate "skin depth" (optical transmission property) for laser scattering technologies [ZHA 02]. The skin depth increases with laser wavelength, ranging from 9 to 80 μm within the wavelengths of 633 and 840 nm. The relative depth of a SSD can be approximated by composing the scatter intensity ratio in the region of the defect to that of the "perfect" material. Thus, by measuring the type, degree, size and shape of a defect's laser scatter signature, a general description of the defect can be derived, along with its location in the sample [STO 90].

7.3.3.2. Applications of laser scattering

Sun *et al.* have built a two-detector laser scattering system to detect SSD in advanced ceramic materials, illustrated in Figure 7.11 [SUN 99].

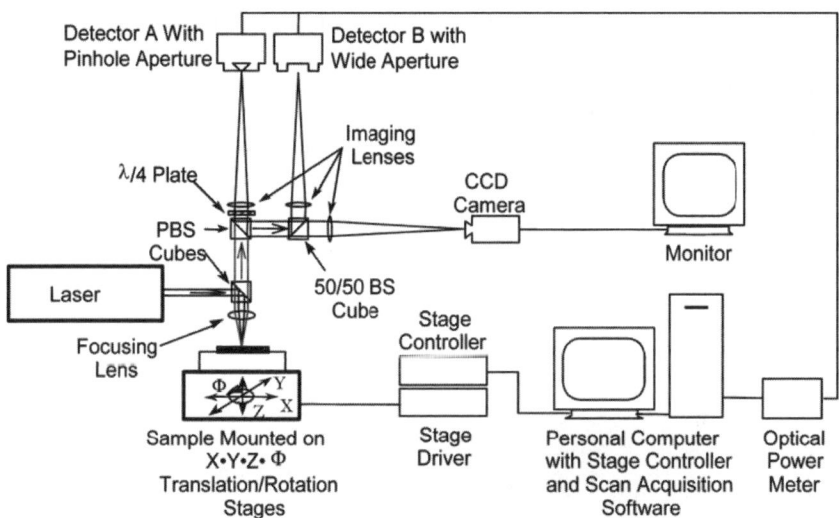

Figure 7.11. *Illustration of a two-detector laser scattering system (after [SUN 99])*

The vertically polarized laser beam was directed through a polarizing beam-splitter (PBS) cube and focused on the sample surface. Unless the surface was extremely rough, the scattered/reflected light from the surface would not change its polarization. Therefore, all the surface-scattered light would be reflected in the PBS and directed back towards the laser. However, any light scattered from the subsurface material underwent several reflections and refractions at microstructural discontinuities (such as cracks), so the subsurface scattered light became depolarized. Half of the subsurface back-scattered light would be reflected by the PBS and directed back to the laser, while the other half would be transmitted by the PBS into the detection train. The horizontally polarized back-scattered light that passed through the surface-illuminating PBS entered into the second PBS (detecting PBS). It was then directed through a quarter-wave ($\lambda/4$) plate, imaged by a positive lens onto a polished stainless steel pinhole aperture and recorded by Detector A. Only the light scattered from the subsurface directly beneath the incident spot passed through the aperture and onto Detector A. The remaining light scattered from the area around the illuminating spot was reflected back through the lens and $\lambda/4$ plate. In this case, its polarization was rotated to vertical and it was reflected by the detecting PBS and directed to a 50/50 beam splitter. One side of this splitter was imaged by a positive lens onto Detector B, while the other side was imaged onto a CCD array to monitor the scattering pattern.

Zhang and Sun used this system to successfully characterize the SSD of three ground silicon wafers that have different SSD depths [ZHA 03, ZHA 05]. These three silicon wafers, identified as wafers A, B and C, were machined by diamond grinding to generate subsurface damage of different depths. First they were coarse-ground by a grinding wheel with #320-mesh diamond grits. Wafers B and C were then fine-ground by a #2000-mesh diamond grinding wheel to remove a surface layer of thickness 10 µm and 30 µm, respectively. Therefore, their estimated SSD depth will be approximately 23 µm, 13 µm and 1 µm respectively. A scattering image of wafer B is shown in Figure 7.12 as an example. The white speckles and lines represent subsurface regions with excessive scattered light due to machining damage.

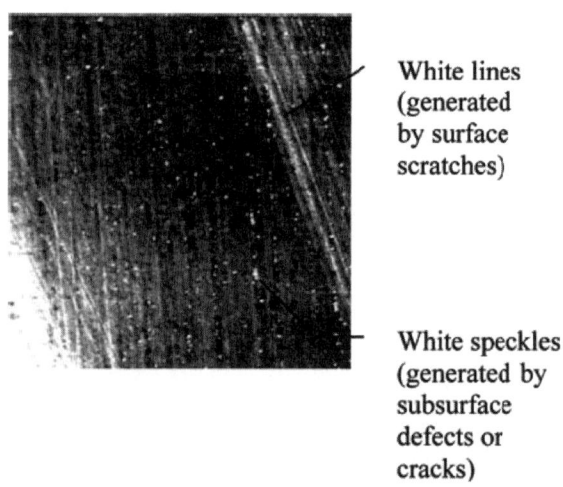

White lines (generated by surface scratches)

White speckles (generated by subsurface defects or cracks)

Figure 7.12. *Laser scattering image of wafer B (after [ZHA 05])*

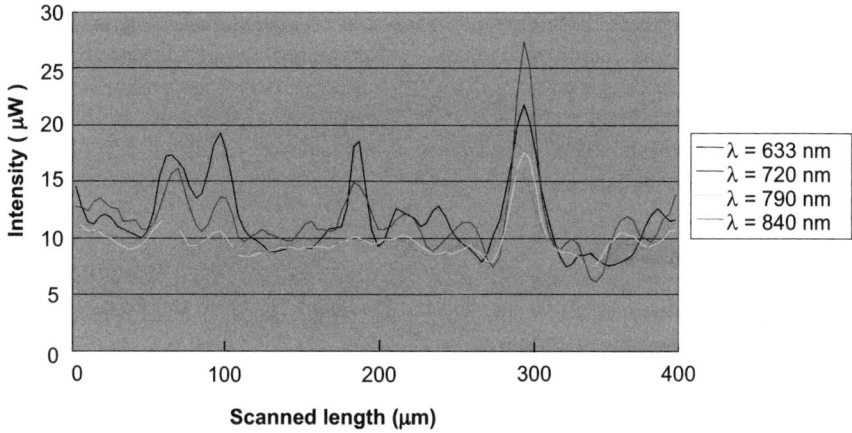

Figure 7.13. *Scatter intensity variation plot for wafer A (after [ZHA 05])*

Proceeding from the scattering image, Zhang and Sun analyzed the scatter intensity variation for each silicon wafer at different laser wavelengths. Because scatter data at one wavelength will only show SSD within the detection depth of that wavelength, shorter wavelengths can only detect shallow damage while longer wavelengths should detect both shallow and deep damage. When different wavelengths are used to scan the same sample, the results can be compared to derive quantitative information about the SSD depth. If SSD shows stronger scatter intensity at the longer wavelength, then the SSD depth should be deeper than the detection depth of the shorter wavelength. On the other hand, if a SSD is found to have higher scatter intensity at the shorter wavelength, the SSD depth should be shallower than or comparable to the detection depth of the shorter wavelength.

As an example, the intensity variation plotted for wafer A is shown in Figure 7.13. It is found that the subsurface scatter intensity at 720 nm wavelength is generally high, and one peak is well above all peaks observed at other wavelengths. This indicates that the depth of the detected SSD is likely to approach the detection depth for 720 nm wavelength, at about 22 µm. At longer wavelengths of 790 nm and 840 nm, the scatter intensities are almost the same for the same scanned position, which suggests that the longer wavelength of 840 nm does not "find" deeper subsurface damage. As expected, the deepest subsurface damage depth for wafer A is about 23 µm. A detailed description and discussion for wafer A and the other two wafers can be found in [ZHA 05]. From these scatter intensity variation plots, it is found that laser scattering technology could identify the location of the deepest subsurface damage on the silicon wafers (where the highest intensity peak occurs), and quantitatively determine the SSD depth on silicon wafers, when combined with the "skin depth" under different laser wavelengths.

7.4. Acknowledgements

This work was supported by the National Science Foundation through the grant CMII-0521203, and the Laboratory Directed Research and Development grant 2004-014-R1 from the Argonne National Laboratory.

7.5. References

[BIS 94] BISMAYER U., BRINKSMEIER E., GUTTLER B., SEIBT H., MENZ C., "Measurement of subsurface damage in silicon wafers", *Precision Engineering*, Vol. 16, 1994, p. 139-143.

[DEG 08] DEGARMO E.P., BLACK J.T., KOHSER R.A., *Materials and Processes in Manufacturing*, 10th Edition, John Wiley & Sons, Inc., 2008.

[GOG 99] GOGOTSI Y., BAEK C., KIRSCHT F., "Raman microscpectroscopy study of processing-induced phase transformations and residual stress in silicon", *Semiconductor Science and Technology*, Vol. 14, No. 10, 1999, p. 936–944.

[HAD 85] HADAMOVSKY H.F., "Mechanische Kristallbearbeitung, in Werkstoffe der Halbleitertechnik", Leipzig: Dt. Verlagf. Grundstoffindustrie, 1985, p. 78-89.

[HAR 04] HARRIS S.J., O'NEILL A.E., YANG W., GUSTAFSON P., BOILEAU J., WEBER W.H., MAJUMDAR B., GHOSH S., "Measurement of the state of stress in silicon with micro-Raman spectroscopy", *Journal of Applied Physics*, Vol. 96, No. 12, 2004, p. 7195–7201.

[HEC 02] HECHT E., *Optics*, 4th Edition, Addison Wesley, 2002.

[JEO 00] JEONG S.M., PARK S.E., OH H.S., LEE H.L., "Fracture strength evaluation of semiconductor silicon wafering process induced damage", *Proceedings of the Fifteenth Annual Meeting of American Society for Precision Engineering*, Scottsdale, AZ, 22-27 October 2000, p. 119-123.

[MCH 95] MCHEDLIDZE T.R., YONENAGA I., SUMINO K., "Subsurface damage in single diamond tool machined Si wafers", *Materials Science Forum*, No. 196-201, 1995, p. 1841-1846.

[MIK 02] MIKULÍK P., BAUMBACH T., KORYTÁR D., PERNOT P., LÜBBERT D., HELFEN L., LANDESBERGER C., "Advanced X-ray diffraction imaging techniques for semiconductor wafer characterization", *Materials Structure*, Vol. 9, No. 2, 2002, p.87–88.

[MIL 02] MILITA S., TIEC Y.L., PERNOT E., CIOCCIO L.D., HARTWIG J., BARUCHEL J., SERVIDORI M., LETERTRE F., "X-ray diffraction imaging investigation of silicon carbide on insulator structures", *Applied Physics A: Materials Science and Processing*, Vol. 75, No. 5, 2002, p. 621–627.

[ONL 00] Online: Limits of resolution: X-ray diffraction, August 2000, http://physics.bu.edu/py106/notes.

[PEI 99] PEI Z.J., BILLINGSLEY S.R., MIURA S., "Grinding-induced subsurface cracks in silicon wafers", *International Journal of Machine Tools and Manufacture*, Vol. 39, No. 7, 1999, p. 1103-1116.

[POP 01] POPOV V.P., ANTONOVA I.V., BAK-MISIUK, J., DOMAGALA J., "Defect transformation study in silicon-on-insulator structures by high-resolution X-ray diffraction", *Material Science in Semiconductor Processing*, Vol. 4, No. 1–3, 2001, p. 35–37.

[QUI 01] QUIRK M., SERDA J., *Semiconductor Manufacturing Technology*, Upper Asddle River, NJ, Prentice Hall, 2001.

[SCH 79] SCHIMMEL D.G., "Defect etch for 100-direction silicon evaluation", *Journal of the Electrochemical Society*, Vol. 126, No. 3, 1979, p. 479-483.

[SPA 88] SPARKS R.G., PAESLER M.A., "Micro-Raman analysis of stress in machined silicon and Germanium", *Precision Engineering*, Vol. 10, No. 4, 1988, p. 191-198.

[SPA 93] SPARKS R.G., PAESLER M.A., "Depth profiling of residual stress along interrupted test cuts in machined germanium crystals", Gaithersburg, MD: NIST Special Publication, No. 847, edited by Jahanmir S, 1993, p. 303-315.

[STE 86] STEPHENS, A.E., "Technique for measuring the depth and distribution of damage in silicon slices", Extended Abstract: Electrochemical Society Meeting, Boston, MA, 4–9 May 1986.

[STE 94] STECKENRIDER J.S., ELLINGSON W.A., "Surface and subsurface defect detection in Si_3N_4 components by laser scattering", *Review of Progress in Quantitative Nondestructive Evaluation*, No. 13, 1994, p. 1645-1651.

[STO 90] STOVER J.C., *Optical Scattering: Measurement and Analysis*, New York, McGraw-Hill, 1990.

[SUN 99] SUN J.G., ELLINGSON W.A., STECHENRIDER J.S., AHUJA S., "Application of optical scattering methods to detect damage in ceramics", in *Machining of Ceramics and Components*, edited by Jahanmir S, Ramulu M, Koshy P., New York, Marcel Dekker, 1999.

[TON 90] TONSHOFF H.K., SCHMIEDEN W.V., INADAKI I., KONIG W., SPUR G., "Abrasive machining of silicon", *Annals of the CIRP*, Vol. 39, No. 2, 1990, p. 621-630.

[TON 94] TONSHOFF H.K., HARTMANN M., PRZYWARA R., KLEIN M., "Defects in silicon wafers: characterization and prevention", *Advancement of Intelligent Production*, 1994, p. 627-632.

[TON 97] TONSHOFF H.K., KARPUSCHEWSKI B., HARTMANN M., SPENGLER C., "Grinding and slicing techniques as an advanced technology for silicon wafer slicing", *Machining Science and Technology*, Vol. 1, No. 1, 1997, p. 33-47.

[WAN 00] WANG S.H., QUAN C.G., TAY C.J., SHANG H.M., "Surface roughness measurement in the submicrometer range using laser scattering", *Optics Engineering*, Vol. 39, No. 6, 2000, p. 1597–1601.

[WOL 96] WOLF I.D., "Micro-Raman spectroscopy to study local mechanical stress in silicon integrated circuit", *Semiconductor Science and Technology*, Vol. 11, No. 2, 1996, p. 139–154.

[YAN 84] YANG K.H., "An etch for delineation of defects in silicon", *Journal Electromechanical Society, Solid State Science and Technology*, Vol. 131, No. 5, 1984, p. 1140-1145.

[YAN 04] YAN J.W., "Laser micro-Raman spectroscopy of single-point diamond machined silicon substrate", *Journal of Applied Physics*, Vol. 95, No. 4, 2004, p. 2094-2100.

[ZHA 02] ZHANG J.M., SUN J.G., PEI Z.J., "Subsurface damage measurement in silicon wafers by laser scattering", *Thirtieth North American Manufacturing Research Conference*, Lafayette, IN, 22-24 May 2002.

[ZHA 03] ZHANG J.M., SUN J.G., PEI Z.J., "Application of laser scattering on detection of subsurface damage in silicon wafers", *Proceedings of International Mechanical Engineering Congress and R&D Expo.*, American Society of Manufacturing Engineering, Washington D.C., 18-21 November 2003.

[ZHA 05] ZHANG J.M., SUN J.G., "Quantitative assessment of subsurface damage depth in silicon wafers based on optical transmission properties", *International Journal of Manufacturing Technology and Management*, Vol. 7, No. 5–6, 2005, p. 540–552.

Chapter 8

Applications of Nano and Micromachining in Industry

8.1. Introduction

High form accuracy and low surface roughness of mechanical, optical and optoelectronic components can significantly improve the quality, the range and ability of functions, and increases the intrinsic value of the final products. Precision and ultra-precision machining, also termed nano and micromachining, is a new technological field which enables production of the high accuracy components, thus is becoming tremendously important technologically and economically in modern society. Micro/nanomachined components are being applied in all kinds of electrical and electronic devices, from high-performance computers for complicated scientific research to cellular phones, digital audio/video recorders and players as household utensils. Micro/nanomachining has become a subject of concentrated research interests from the manufacturing industries of various high-tech products.

In this chapter, micro/nanomachining technologies for metal materials, semiconductor materials and single-crystalline materials will be introduced and discussed based on the recent research outcomes of the author's research group in collaboration with the manufacturing industry. A few examples of the application micro/nanomachining technologies to the fabrication of liquid crystal display (LCD) components, infrared aspheric optics, Fresnel lenses and semiconductor substrates will be highlighted.

Chapter written by Jiwang YAN.

8.2. Typical machining methods

In order to realize micro/nanomachining, it is essential to realize an extremely small machining scale. In 2D machining, the machining scale is usually called "undeformed chip thickness", and in 3D machining, it is called "cut depth". The micro/nano level machining scale is guaranteed by extremely high-precision and high-stiffness machine tools and related measurement and control technology.

There are numerous machining methods, according to the types of tools used for machining. Generally, mechanical machining can be divided into two large categories: cutting and abrasive machining. Cutting can be further divided into single-point cutting and multiple-point cutting. Milling and drilling are two typical multi-point cutting methods. In this chapter, however, only single-point cutting technology is highlighted. Furthermore, according to the size and shape of required components, the machine tools used for micro/nano-cutting are designed or various different structural types. Here we briefly introduce two of the most commonly used single-point cutting methods: diamond turning and shaper/planner machining.

8.2.1. *Diamond turning*

Of the many machining methods, diamond turning is the most popular method of performing nano-cutting. Diamond turning is a process of mechanical fabrication for precision elements using digitally controlled lathes equipped with natural or synthetic diamond-tipped cutting tools. This technology is not new, but emerged in the 1960s with the manufacturing demands in science and technology for energy, computer, electronics and defense [WHI 66, BRY 67, KRA 69, IKA 91]. Due to the fact that the machining scale is extremely small compared to conventional machining, it is also called single-point diamond turning (SPDT). SPDT has now become a well-established technology for the fabrication of non-ferrous metal mirrors for optical applications.

Some of the features that enable SPDT technology are: extremely sharpened diamond cutting tools, air-bearing spindles, pneumatic/hydrostatic slides, feedback control, vibration isolation, and temperature control, etc. The lathe used for SPDT usually rests on a high-quality granite base with fine surface quality. The granite base is placed on air suspensions on a specially constructed solid foundation, keeping its working surface strictly horizontal. The machine tables are placed on the top of the granite base and can be moved with a high degree of accuracy using high-pressure aerostatic or hydraulic suspensions with high stiffness. The machine elements are usually driven by servomotors via hydrostatic screws or directly driven by high speed linear motors. The movement of the machine elements is measured and monitored by extremely high-resolution displacement sensors such as laser hologram scales. The workpiece is usually attached to a vacuum chuck using negative air pressure after being strictly centered, and then rotated at a high speed by an aerostatic or hydraulic bearing spindle.

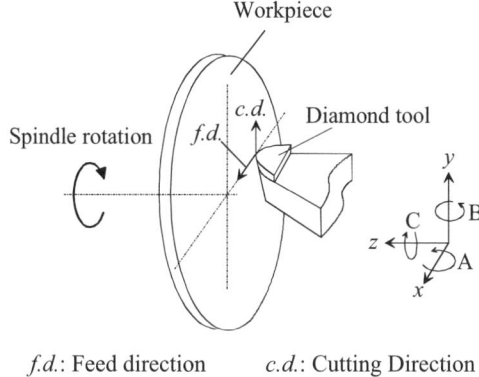

f.d.: Feed direction   c.d.: Cutting Direction

Figure 8.1. *Schematic presentation of diamond turning*

Figure 8.2. *Photograph of the main section of a diamond turning machine*

A schematic presentation of the diamond turning operation is shown in Figure 8.1, and a photograph of the main section of a diamond turning machine is shown in Figure 8.2. The material removal in SPDT is realized by the relative movement between the workpiece and the cutting tool, that is, the linear movement of the machine tables and the rotation of the workpiece. The change of the machining scale, i.e. the undeformed chip thickness, is realized by changing the tool feed rate and the cut depth. Recent ultra-precision machines have already enabled four- or five-axes numerical control at a resolution better than one nanometer per step to fabricate nano-precision components with very complicated shapes. The development of this manufacturing technology has been a revolutionary advancement of the conventional machining technology for extreme precision.

8.2.2. Shaper/planner machining

Shaper and planner are also popular single-point precision machining methods. A shaper operates by moving a cutting tool backwards and forwards across the workpiece. The workpiece mounts on a rigid square table that can traverse sideways underneath the reciprocating tool, which is mounted on a ram. The table motion is usually controlled by a precise feed mechanism. The ram slides back and forth above the workpiece and the tool can be positioned to cut the flat surface on the top of the workpiece. A planer is a type of machining tool analogous to a shaper, but larger and with the entire workpiece moving beneath the cutter instead of the cutter moving above a stationary workpiece. A schematic presentation of shaper and the planner machining is shown in Figure 8.3.

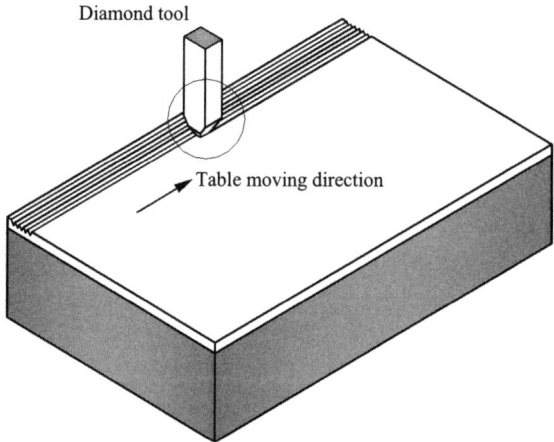

Figure 8.3. *Schematic presentation of shaper/planner machining*

Planers and shapers have generally been used for two types of work: generating large, accurate, flat surfaces and cutting straight microgrooves. Modern planers are used for producing precision stamping dies and plastic injection molds for light-guiding plates of liquid crystal displays, large-scale linear Fresnel lenses, etc. Recently developed shaper machines have also made the fabrication of curved microgrooves and 3D microstructures possible, by controlling the orientation angle and moving direction of the cutting tool and the cut depths.

8.3. Applications in optical manufacturing

Household utensils such as cellular phones, digital audio and video players are becoming more and more functional, and simultaneously more and more compact. This advance is partially due to the miniaturization of the optical systems inside these products. Integration of multiple functions and miniaturization in size has been achieved by introducing advanced optical elements which have complex shape and high accuracy. Examples of the advanced optical elements are aspheric lenses, Fresnel lenses, diffraction optical elements (DOE) and other hybrid optical components.

8.3.1. *Aspheric lens*

An aspheric surface is expressed as the difference between a sphere and an asphere at different heights above the optic axis [SHE 85]. An axisymmetric aspheric surface can generally be described by

$$z(x) = \frac{Cx^2}{1 + \sqrt{1 - (k+1)C^2 x^2}} + \sum_{i=1}^{m} a_i x^i \qquad [8.1]$$

where $C=1/r$, in which r is the radius of curvature for the spherical surface; x is the distance from the optic axis (Z); k is the conic constant, a parameter representing the eccentricity of the conic surface; a_i for even i are the aspheric deformation constants, and a_i for odd i are aspheric coefficients used to define other polynomial curves by setting $C=0$.

Aspheric lenses can, remarkably, eliminate spherical aberrations on a spherical lens. The function of a large group of spherical lenses can be replaced by a single aspheric lens, which enables miniaturization and low-cost production of optical systems. Most optical lenses used in the visible light wavelength are made of glass and plastics. Glass lens has many predominant advantages over the plastic counterpart in terms of its hardness, refractive index, light permeability, resistance to environmental changes (in terms of temperature and humidity), etc. For this reason, glass lenses have been needed increasingly in the field of high-resolution digital cameras, mobile phone cameras and CD/DVD players. Conventionally, glass aspheric lenses are fabricated by a series of material removal processes [NIC 81, JOH 05] such as grinding, lapping and polishing, which requires a long production cycle and results in a very high production cost. As an alternative approach, the glass molding press (GMP) process has been shown to be a promising way to efficiently produce precision aspheric lenses [KAT 06, MAS 07]. Figure 8.4 is an example of a glass aspheric lens produced by glass molding technology.

A key technology for manufacturing aspheric lenses is mold fabrication. Molds for aspheric plastic lenses are usually made of metals such as copper or stainless steel plated with nickel-phosphorus alloy (NiP), by diamond turning. Molds for glass lenses are conventionally made of cemented carbide or silicon carbide, which must be ground and polished. Recently, we have demonstrated that specially designed NiP plated metal molds can also be used for glass molding [MAS 07]. The mold plated with NiP can be processed by nano and microcutting with a single crystal diamond tool to obtain extremely fine surface [YAN 04-1].

Figure 8.4. *Example of an aspheric lens*

The light with a longer wavelength than that of the visible light is infrared light. Today, infrared optical lenses are increasingly required in various industrial fields, such as night vision systems for vehicles, dark field sensing systems for security, etc. Infrared optical lenses are usually made of brittle materials such as single crystalline germanium and silicon instead of glass. Silicon is an important optical material commonly used in infrared imaging systems as lenses, windows, mirrors and output couplers for low power lasers [KAP02]. Germanium has very high permeability and high refractive index in the wavelength range of 2~14 microns; thus, it has been an excellent substrate material for infrared lenses with extensive applications in thermal imaging systems, dark field optical instruments, astronomical telescopes, night-vision systems, etc. [TEE 99, CHA 94].

Although new infrared transparent materials other than silicon and germanium are being developed and press molding technology has been used to fabricate infrared lenses at a relatively low production cost, these new materials are still expensive. Generally, the newly developed materials must be molded at a very high temperature; thus, the life of the molding die is very limited, and in turn, the expense of the pressing molds is very high. For these reasons, press molding technology has not yet been widely used in the infrared optical manufacturing industry. Currently the fabrication of infrared lenses is still dependent on material removal processes.

Conventional infrared lenses are of spherical shape and are manufactured by grinding and polishing. Recently, aspheric lenses have been in increasing demand, and these complicatedly shaped lenses are mainly manufactured by diamond turning. Studies have revealed that plastic deformation occurs to silicon and germanium without brittle fracture during micromachining tests, and that the plastic deformation is facilitated by the high-pressure phase transformations [NAK 90, BLA 90, YU 94, MOR 95, YAN 98, SYN 98, LEU 98]. These findings provided insights into the fundamental physics governing the micro-deformation mechanism and contributed significantly to the ductile machining technology of brittle materials. In previous work, we investigated the ductile machinability of silicon and germanium with various crystal orientations. A crack-free surface with nanometer level surface roughness can be obtained by maintaining the machining mode as a ductile [YU 94, YAN 98, YAN 02-1, YAN 04-2, YAN 06-1].

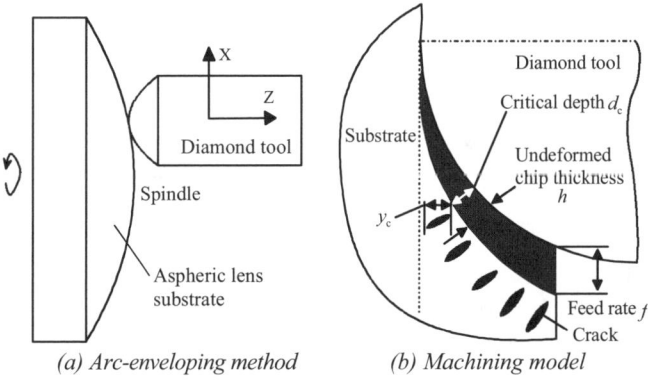

(a) Arc-enveloping method   (b) Machining model

Figure 8.5. *Schematic models of the aspheric surface cutting with round tools*

In the ductile machining of silicon and germanium lenses, a small radius round-nosed diamond tool with high form accuracy has been used under 2-axis digital control, as shown in Figure 8.5(a). The tool radius is usually about 1 mm or much smaller [NAK 90, BLA 90, LEU 98]. As shown in Figure 8.5(b), a truly ductile response only occurs along the apex of the tool tip where the undeformed chip thickness is smaller than a critical value (critical depth d_c), while the upper material is fractured. To obtain a crack-free surface using this kind of tool, the tool feed f must be extremely small (1~3 μm/rev). As a result, ductile machining by this method is inefficient with a very low material removal rate. In particular, when machining a large-diameter component tool, wear will become a critical problem.

In order to conduct an efficient ductile machining, diamond turning of brittle materials using a straight-nosed diamond tool was proposed [YAN 98, YAN 02-1]. This has been successfully used in the fabrication of convex aspheric lenses on single crystalline silicon and germanium on 3-axis digitally controlled machining tools [YAN 02-2]. This method was termed the straight-line enveloping method (SLEM). A schematic representation of the method is shown in Figure 8.6(a). The tool is moved in the X and Z directions and simultaneously rotated about a B-axis that is perpendicular to the X-Z plane. The objective aspheric surface is then enveloped by the straight edge of the tool. The corresponding cutting model is shown in Figure 8.6(b). In this case, undeformed chip thickness h is uniform within the entire cutting region and is determined by tool feed f and cutting edge angle κ. Therefore, by using a sufficiently small cutting edge angle κ, an extremely small undeformed chip thickness h can be obtained even at a large tool feed f. Ductile machining at a large tool feed improves both machining efficiency and tool life. In addition, the straight-nosed tool used in this method is easier to manufacture than the round-nosed tool, thus low-cost production can be expected.

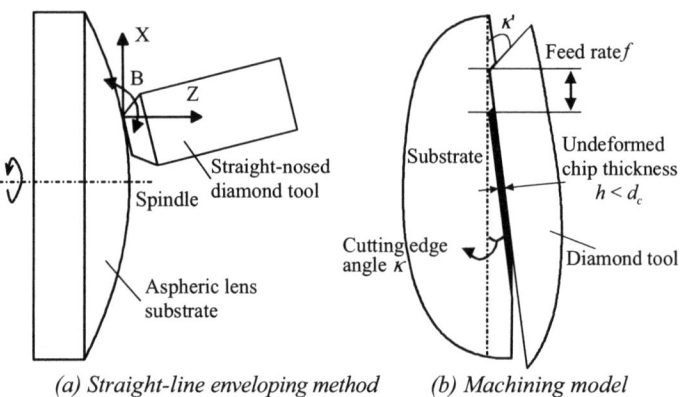

(a) Straight-line enveloping method   (b) Machining model

Figure 8.6. *Schematic models of aspheric surface cutting by straight tools*

Applications of Nano and Micromachining in Industry   183

Figure 8.7. *Photograph of the main section of the machine tool*

As can be seen in Figure 8.6(a), the SLEM method requires the use of a 3-axis (*X-Z-B*) machine tool and these axes must be simultaneous numerically controlled. Figure 8.7 is a photograph of the main section of a 3-axis (*X-Z-B*) simultaneous numerically controlled machine tool. This machine has an air bearing spindle with a maximum rotation rate of 2,500 rpm, two perpendicular hydrostatic slide tables along the *X*-axis and *Z*-axis respectively as well as a hydrostatic rotary table around the *B*-axis. The hydrostatic bearing has the advantages of high stiffness and low friction. The motions of all three axes are real-time measured by linear laser scales and numerically controlled in a closed loop. This provides the *X* and *Z* slide tables with a linear motion revolution of 10 nm per step, and the *B*-table with an angular motion resolution of 0.001°, respectively. The machine base was made of granite and supported by air mounts for the purpose of isolating external vibration.

184   Nano and Micromachining

Figure 8.8. *(a) Photograph of a silicon aspheric lens fabricated by diamond turning with a straight tool, (b) SEM photograph of cutting chips generated during ductile machining of silicon*

Figure 8.8(a) is a photograph of a silicon aspheric lens fabricated by diamond turning with the SLEM method at an undeformed chip thickness of 84 nm. The entire surface is a mirror-like smooth surface with no damaged area observed. This large-diameter lens could be fabricated in a few minutes by using a high tool feed rate up to 40 μm per revolution of spindle. Figure 8.8(b) is an SEM photograph of cutting chips generated during ductile machining of silicon. The chips are long continuous ribbons, indicating a complete ductile machining mode.

Applications of Nano and Micromachining in Industry 185

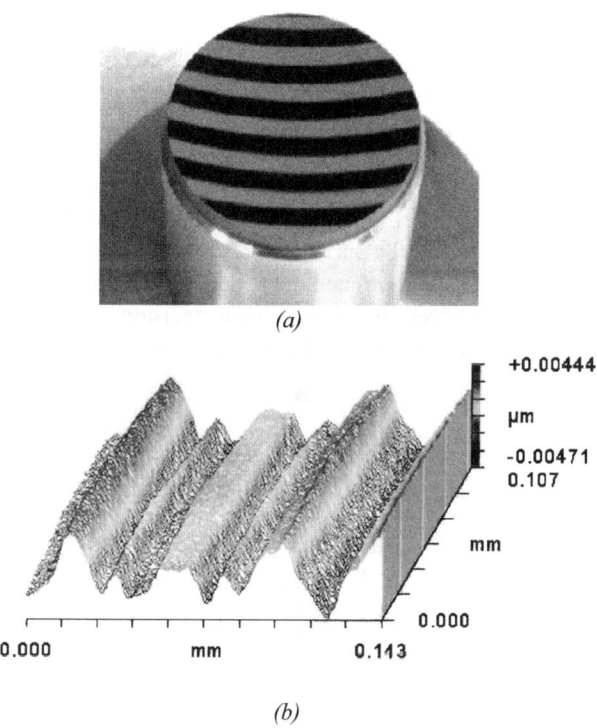

Figure 8.9. *(a) Photograph of a germanium aspheric lens fabricated by diamond turning with a large-radius tool, (b) surface roughness profile*

However, for straight-nosed tools, it is difficult to generate concave surfaces because of the interference between the tool and the workpiece. In a recent work, we attempted to conduct high-efficiency ductile machining of germanium with a large radius round-nosed diamond tool to solve the aforementioned problems. The tool nose radius used in the present experiments ranges from a few millimeters to a few tens of millimeters, far larger than that of the conventional tools (~1 mm). We used a tool feed rate 5 times higher than the conventional rate and a spindle rotation rate 3 times the conventional rate (~5,000 rpm). Theoretically, these conditions give rise to a material removal rate 15 times higher than that of the conventional process. It has been demonstrated that this method can achieve both high production efficiency and high surface quality [OHT 07]. Figure 8.9(a) is a photograph of an aspheric germanium lens fabricated by diamond turning with a large-radius (10 mm) tool. The lens surface is mirror-like and extremely smooth. The surface roughness is below 10 nmRy, as shown in Figure 8.9(b), and the form error is smaller than 0.2 μm.

8.3.2. Fresnel lens

Recently, the demand for Fresnel lenses has become increasingly prominent. A Fresnel lens is a plano-convex or plano-concave lens which has been cut into concentric narrow rings, namely Fresnel zones, and flattened. It retains the optical characteristics of a plano-convex or plano-concave lens but is much smaller in thickness, and therefore has less absorption losses. There are numerous optical designs that can benefit from the application of the Fresnel lenses. If the Fresnel zones are sufficiently narrow, the surface of each zone can be made approximately conical (with straight cross section) and not spherical (with arc cross section) for the ease of production. However, for recent high-precision optical systems, Fresnel lenses including spherical zones with arc cross-sections are necessary.

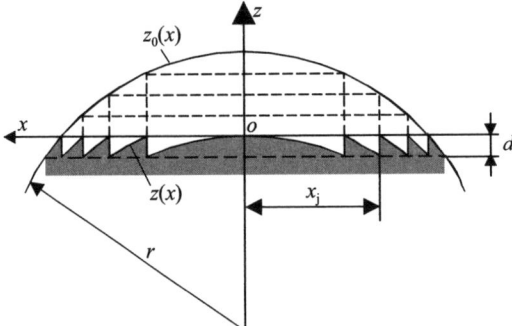

Figure 8.10. *Schematic of the geometry of a Fresnel lens*

Figure 8.10 shows the geometry of a Fresnel lens. The shape of the Fresnel structure, $z(x)$, can be described by the modulus function (MOD) of the equivalent plano-concave or plano-convex lens shape $z_0(x)$ and the zone depth d as:

$$z(x) = MOD[z_0(x), d] \qquad [8.2]$$

where the MOD function is a function that returns the remainder when one number is divided by another. For example, MOD (4, 3) = 1.

The surface function of a spherical lens shape $z_0(x)$ can be simplified from that of an aspheric surface. In optical design and manufacturing, the topology of an axis-symmetric aspheric surface is usually expressed as equation [8.1], as mentioned in section 8.3.1. For a spherical shape, we can just set $k=0$ and $a_i=0$ in equation [8.1]. Thus, the function of a spherical lens surface can be described as:

$$z_0(x) = \frac{Cx^2}{1 + \sqrt{1 - C^2 x^2}} \qquad [8.3]$$

This equivalent lens surface is cut into concentric rings by cylindrical surfaces at the zone steps. The radial coordinate of each zone step can be calculated by:

$$x_j = \sqrt{j \cdot d \left(\frac{2}{C} - j \cdot d\right)} \qquad [8.4]$$

where j is the sequential number of the Fresnel zone counted from the optical axis z, and d is the zone depth, as in equation [8.2].

Photolithography and diamond turning are two major machining methods used for fabricating micro-Fresnel structures. Lithography technology is normally used to produce 2D microstructures on flat substrates. By using a varying dose during the illumination, it is also possible to fabricate 3D microstructures such as Fresnel and blazed diffractive optical elements using recent lithography technology. However, the structural depth is limited by using lithography techniques. Especially for Fresnel structures with large depth, the low material removal efficiency of lithography becomes a problem. For these applications, diamond turning will be preferable [YAN 05-1].

Figure 8.11. *Schematic presentation of machining method for Fresnel lens*

Figure 8.11 shows the machining method for Fresnel lenses. The Fresnel structure is generated on a flat substrate using a diamond tool which has a sharply-pointed V-shaped tip. Figure 8.12 is an SEM micrograph of a diamond tool used for cutting Fresnel lenses. The tool is made from a single-crystal diamond and has a 60° included angle and an extremely sharpened V-shaped tip which theoretically has no nose radius. The roundness of the tool tip was estimated to be smaller than 1 micron by SEM observation.

188   Nano and Micromachining

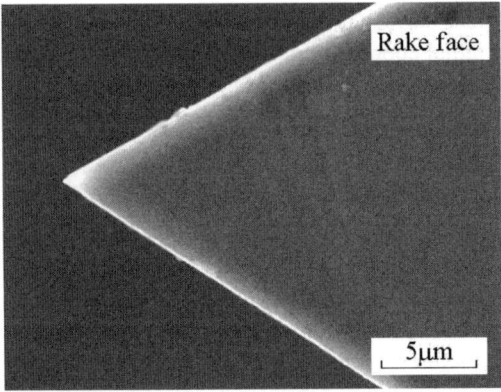

Figure 8.12. *Scanning electron micrographs of the diamond cutting tool*

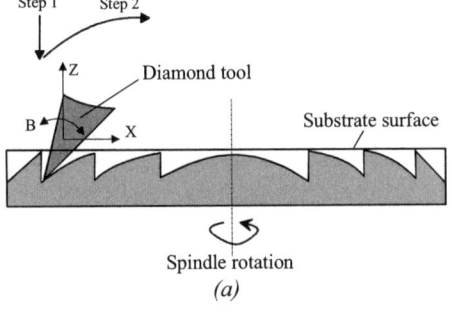

Figure 8.13. *Machining models for the Fresnel structures: (a) process steps, (b) model for ductile machining of a single microgroove*

Figure 8.13 shows the machining models for fabricating a Fresnel lens. The machining operation for one micro groove consists of two steps, as shown in Figure 8.13(a). First, the tool moves along the Z-axis, namely the vertical axis, to generate the cylindrical surface at the zone step. Next, the tool moves and at the same time rotates under X-Z-B three-axis simultaneous control to generate the spherical surface. The detailed schematic model of step 2 is shown in Figure 8.13(b). In the figure, the thickness of the material removed during one revolution of the workpiece, namely the undeformed chip thickness h, is determined by:

$$h = f \cdot \sin \kappa \qquad [8.5]$$

where κ is the angle between the cutting edge and the tangent of the objective curved surface, namely the cutting edge angle, and f is the tool feed rate. From equation [8.5], we can see that by using a small cutting edge angle κ, the undeformed chip thickness h can be decreased without reducing the tool feed rate f. Therefore, if h is thinned to be smaller than the critical undeformed chip thickness, a ductile-cut surface can be obtained.

When fabricating a Fresnel lens, it is important to prevent the edges of the zone from microfracturing. One of the possible reasons for microfracturing is the subsurface damages caused by precuts. Usually, precuts are required to make the lens substrates flat and smooth before microgrooving. However, the precutting process may cause subsurface damages such as potential microcracks and dislocations. If the lens apexes are located within the damage layer, microfractures readily occur to the zone steps. In order to solve this problem, we used a depth-offsetting method, that is, by off-setting the lens apexes a few microns lower than the substrate surface, the precut-induced damage layer will be removed during the finishing cut, without damaging the micro-Fresnel structures.

Machining experiments were carried out on a three-axis digitally controlled ultra-precision lathe, NACHI-ASP15. As a test piece, a Fresnel lens which has a curvature radius of 100 mm was fabricated. Thus, in equation [8.3], $C=0.01$. For practical applications, the zone depth d in equation [8.4] should be decided according to the design wavelength, the refractive index of substrate material and the focus length. In this work, we set the zone depth d to 50 microns to avoid particularity. A single-crystal germanium (110) substrate was used as workpiece. It is optical-grade pure germanium with no doping. The workpiece was 30 mm in diameter, 5 mm in thickness and obtained with ground finish. To remove the damaged layer due to grinding, precuts were performed with other cutting tools, providing a mirror-like flat surface. The position of the Fresnel lens apex was offset to 2 microns lower than the precut surface.

For step 1 in Figure 8.13(a), i.e., when generating the zone steps, the tool feed rate f_1 was 0.05 μm/rev; and during step 2, the feed rate f_2 was set to 15 μm/rev. The cutting edge angle κ in Figure 8.13(b) was set to 0.2°. Under these conditions, the undeformed chip thickness h was approximately 50 nm, which is smaller than the critical undeformed chip thickness for ductile machining at all crystal orientations

(~60 nm). The rotation rate of the machine spindle was fixed to 1,000 rpm. Kerosene mist was used as coolant.

Figure 8.14. *Scanning electron micrographs of the fabricated Fresnel lens*

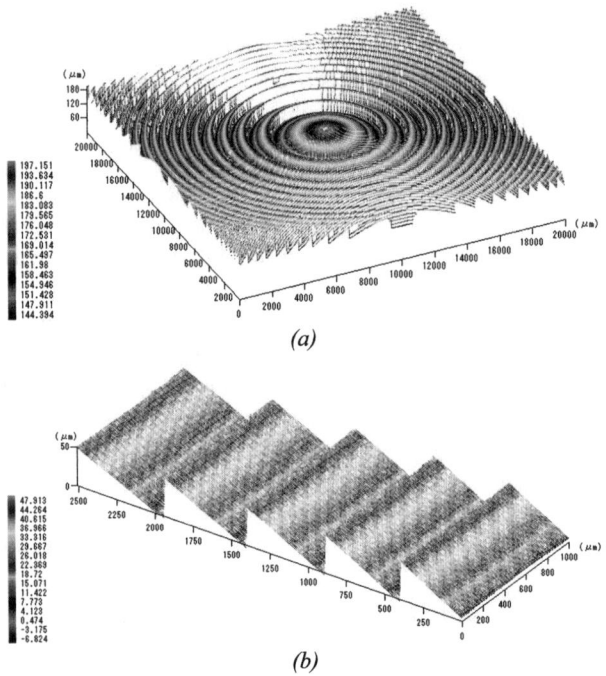

Figure 8.15. *3D topographies of the lens: (a) central region, (b) outer region*

Figure 8.14 is an SEM photograph of the fabricated Fresnel lens. It can be seen that the lens surface is very smooth, without any sign of brittle fractures. The zone steps can be identified clearly as concentric rings, indicating that the microgrooves have been ductile-cut. The Fresnel lens was finished by a single tool pass, and the total time for machining was approximately 30 minutes, demonstrating a high machining efficiency.

A non-contact three-directional measuring machine was used to measure and evaluate the lens geometry. This measuring instrument uses a semiconductor laser probe (wavelength 635 nm) to scan the lens surface, thus avoiding the contact damages due to a conventional stylus profiling instrument. The resolution of this measuring instrument is 1 nm. The laser beam has an extremely small spot size, approximately 1 µm, so that the narrow microgroove corner can be precisely measured without the edge-rounding effect that occurred in traditional stylus-profiling methods. Although sometimes the large incident angle may cause errors at steep rims of microgrooves, the problem can be solved to some extent by selecting suitable measuring conditions and by processing the measured data using software techniques. Figure 8.15 shows the 3D topographies of the lens center and a part of the outer region, respectively. It can be seen that the Fresnel structures have been precisely fabricated both at the workpiece center and at the outer regions, without any visible defects.

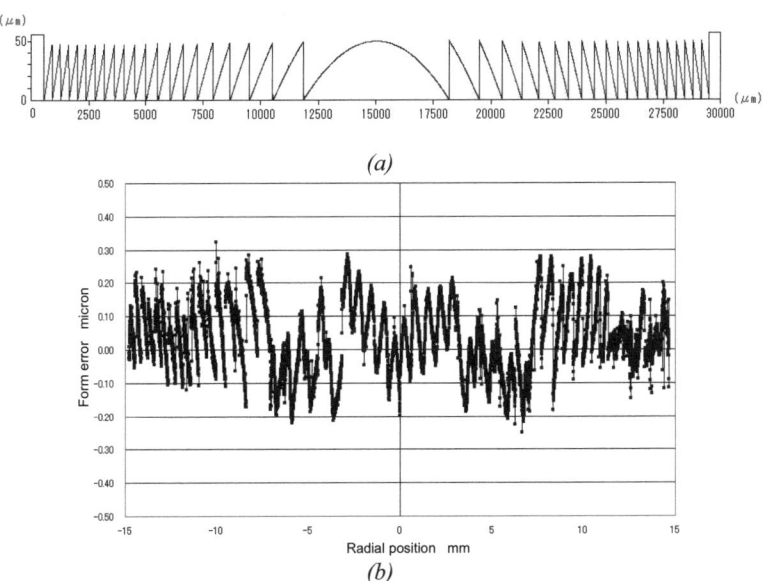

Figure 8.16. *2D measurement result of the lens geometry: (a) cross-sectional profile, (b) form error distribution*

Figure 8.16(a) shows a cross-sectional profile of the Fresnel lens. It can be seen that the depths of all the microgrooves are uniform, without roundness or microfractures at the groove profiles. Figure 8.16(b) shows the form error distribution which is calculated by comparing the measured cross-sectional profile with the designed profile. The form error distribution is almost uniform, with minor random variations. The peak to valley amplitude of the form error is approximately 0.5 micron. The surface roughness of the microgrooves was measured along two directions, namely, the circumferential direction (cutting direction) and the radial direction (feed direction). The average surface roughness along the cutting direction was 20 nmRy and that along the feed direction was 50 nmRy, respectively, where Ry is the peak to valley, or maximum height, of the surface. Since the radial surface roughness is determined by the tool-work transcription principle, a much smoother surface can be obtained by using a smaller tool feed and/or a smaller cutting edge angle.

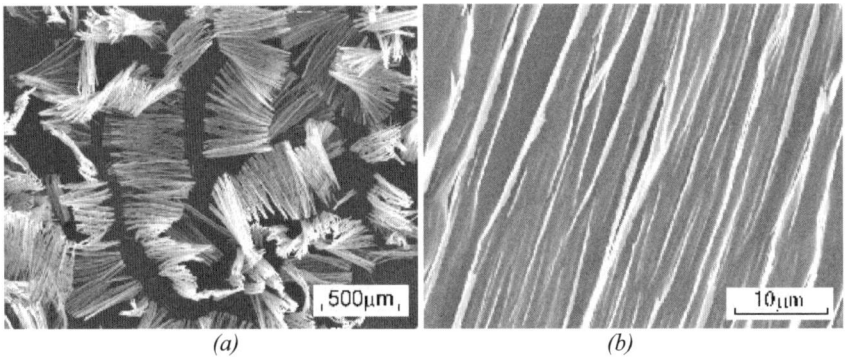

(a) (b)

Figure 8.17. *Scanning electron micrographs of the cutting chips:*
(a) general view, (b) detailed view

Figure 8.17(a) is an SEM photograph of the cutting chips collected during the machining process, while Figure 8.17(b) is a detailed view of the chip surface. The chips are long and generally continuous, showing plastic deformation appearance. These chips indicate that as a nominally brittle material, single-crystal germanium has been subjected to significant plastic deformation and machined in a complete ductile mode.

Other optical crystals such as calcium fluoride (CaF_2) which is the most suitable material for 193 and 157 nm lithography systems, non-linear crystals KDP, $KNbO_3$ and $LiNbO_3$ are also widely used in optical devices. These materials are indispensable in laser frequency conversion, frequency mixing and optical parametric oscillation, etc. The nanomachining technology of these materials is also an important issue for industry. Some advances have been made in nanolevel ductile machining of these materials, but further investigations are needed [YAN 04-3, YAN 04-4].

8.3.3. *Microstructured components*

3D microsurface structures, such as arrays of microgrooves, micropyramids, microlenses and microprisms, are required more and more in recent optical, optoelectronic, mechanical and biomedical industries. Components with microsurface structures yield new functions for light operation, thus significantly improving the imaging quality of optical systems. Microstructures can also be used as fluid channels in biomedical and biochemical applications. Therefore, high-precision and high-efficiency fabrication of the microstructures on flat or curved surfaces are the subject of intensive interest.

Glass and plastic are two major substrate materials for microstructured components. Generally speaking, plastic components are easily manufactured by molding technology at a low production cost. On the other hand, glass has considerable advantages over plastic in terms of hardness, refractive index, light permeability, resistance to environmental changes (in terms of temperature and humidity), etc. Recently, the industrial need for glass components has been on the increase.

A few microstructures on glass can be fabricated by material removal processes such as sand blasting, photolithography and wet/dry etching. These processes are effective for manufacturing microstructures with rectangular cross-sections, but ineffective for microstructures with angled or curved cross sections. Microcutting of glass with micro-endmills has also been reported, but the production efficiency is limited and the production cost is extremely high for mass production.

As an alternative approach, glass molding is a promising method for producing precision optical elements such as aspheric lenses, Fresnel lenses, diffractive optical elements (DOEs), microlens arrays, etc. As mentioned in section 8.3.1, in glass molding, the fabrication of the molding die is an important issue. Although hard materials such as silicon carbide (SiC), tungsten carbide (WC) and fused silica (SiO_2) are preferable mold materials for continuous surfaces, they have not been commonly used for molding microstructures because it is very difficult to generate microstructures on these materials.

Nickel-phosphorous (NiP) electroless plating is known as an important mold surface preparation technology for plastic optical parts manufacturing. NiP plating provides hard, wear-resistant and corrosion-resistant surfaces at relatively high temperatures and, at the same time, maintains excellent precision micromachinability. It has been preliminary demonstrated that NiP plating can be used for molding aspheric and diffractive glass lenses with a long mold life in practice [MAS 07]. In this section, we introduce the fabrication of 3D microstructures on electroless plated NiP substrates by nano and microcutting technology, and highlight some examples of hot-press glass molding tests using the micromachined microstructures as molding dies [YAN 08-1].

Two kinds of microstructures were fabricated by microcutting: one is microgrooves and the other is micropyramids. To cut a microgroove, a sharp diamond cutting tool is transversely fed in the horizontal direction while the cut depth is kept constant. The principle of microgrooving is based on shaper/planner machining as mentioned in section 8.2.1. By periodically shifting the tool perpendicularly to the cutting direction, parallel microgroove arrays can be generated. After the groove arrays have been fabricated, the workpiece is then rotated at an angle of 90° on the horizontal plane, and the cross-grooving operation is performed. In this way, micro-pyramid arrays can be formed on the workpiece surface. Microcutting tests were done on an ultra-precision machine Toyoda AHN-05. The machine enables its tables to move under four-axis (XYZB) numerical control at a stepping resolution of 1 nm. Figure 8.18 is a photograph of the main section of the machine. A piezoelectric dynamometer Kistler 9256A was mounted below the workpiece to measure the microcutting forces during the cutting tests.

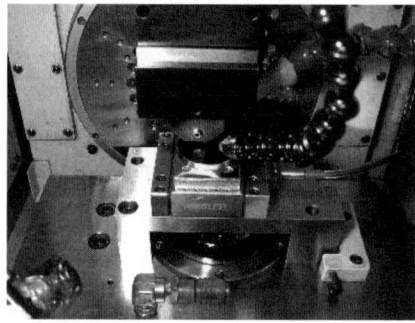

Figure 8.18. *Photograph of the microcutting setup*

Applications of Nano and Micromachining in Industry   195

Figure 8.19. *SEM photograph of the microcutting tool with a micro-chamfered tip*

Two kinds of cutting tools made of single-crystal diamond were used: one is an extremely sharpened tool with no end width, and the other has a 5 μm flat chamfer at the tool tip. Two kinds of included angles were used, namely 90° and 60°, respectively. The rake angle is 0° and the relief angle is 10° for all the tools. Figure 8.19 is an SEM photograph of the tip of a micro-chamfered 60° diamond tool. As a test workpiece, a cylindrical stainless steel blank with a diameter of 50 mm electroless-plated with NiP was used. The thickness of the NiP layer was approximately 100 μm. Diamond turning was performed to flatten the NiP plated surface before machining microstructures. The depth and the pitch of the microgrooves were both changed from 1 μm to 10 μm level. The cutting speed was set to 500 mm/min (0.0083 m/s). As lubricant, the Bluebe #LB10 cutting oil was used in the form of a mist jet.

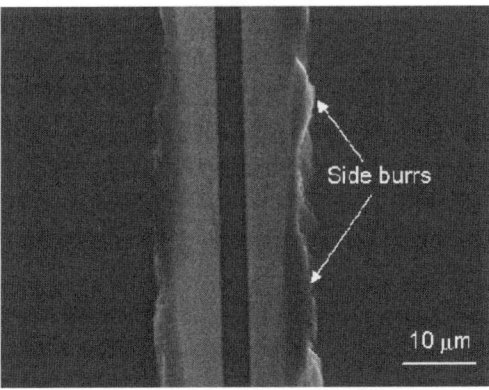

Figure 8.20. *SEM photographs at different magnifications of microgrooves with side burrs*

196   Nano and Micromachining

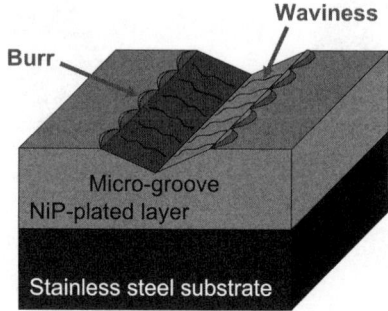

Figure 8.21. *Schematic model of side burrs and waviness in micro grooving process*

Burr formation was found to be a critical problem in microgrooving, especially in dry cutting or when the cutting fluid is not sufficiently provided into the cutting region [YAN 04-1, YAN 08-2]. Figure 8.20 shows an SEM photograph of microgrooves cut by single cuts at a depth of 10 μm with a micro-chamfered 60° tool. It can be seen that burr formation at the side edges of the microgroove is very significant, although the inside surface of the groove is smooth. Within a few microgrooves, surface waviness was also observed at the groove bottom. Figure 8.21 schematically shows the formation model of side burrs and waviness in the microgrooving process.

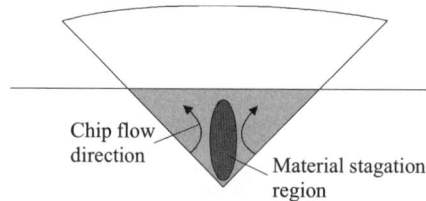

Figure 8.22. *Cross-sectional model of the one-step grooving process*

Applications of Nano and Micromachining in Industry   197

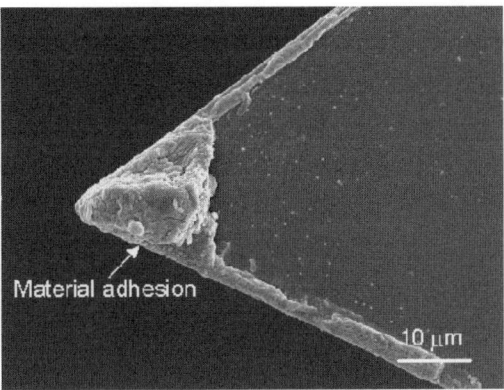

Figure 8.23. *SEM photograph of the tool tip showing material adhesion*

The burr formation is presumably caused by the side flow of material in the cutting region. As shown in Figure 8.22, the cross-section of the uncut chip is triangular, which is distinctly different from normal orthogonal cutting conditions where the uncut chip thickness is uniform along the cutting edge. Therefore, during cutting the material flows generated by the two side edges are directed towards the center of the tool and interfere with each other, leading to a high-pressure material stagnation zone. The material stagnation at the tool center will cause significant side flows of material to form side burrs. At the same time, microscopic waviness may occur due to the unsteadiness of material flow and fluctuation of cutting forces. This problem is especially serious when cutting deep grooves or grooves with high aspect ratios.

To examine the occurrence of material stagnation, SEM observation of the tool tip was performed after cutting. Figure 8.23 is an SEM photograph of the tool tip where material adhesion can be observed. Material stagnation causes a highly negative effective rake angle and, in turn, increases cutting forces. The phenomenon of material stagnation can be partially prevented by applying cutting fluid effectively into the cutting region, but it is difficult to eliminate it completely.

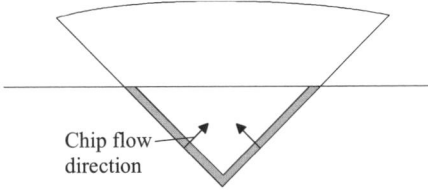

Figure 8.24. *Cross-sectional model of the second cut of the two-step microgrooving process*

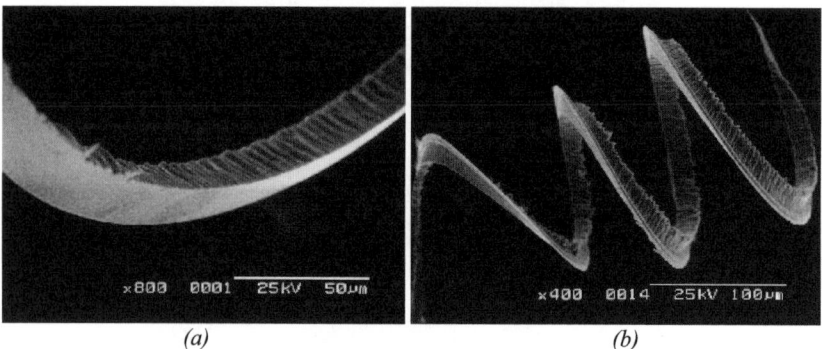

Figure 8.25. *SEM photographs of cutting chips in (a) the first cut and (b) the second cut*

As one of the solutions to the burr formation problem, two-step cutting is effective, that is, the first cut is done as a rough cut at a large depth, and the second cut is done as a finish cut at a small chip thickness. In this case, the cutting model of the finish cut is shown schematically in Figure 8.24. The cutting conditions at both side edges are similar to the orthogonal cutting state. This way, the material flow will be fluent, which contributes to burr-free surface formation.

Figure 8.25 is a comparison of chip geometry. In the first cut of the two-step cutting, a triangular cross-section chip is formed; whereas in the second cut, the chip has a V-shaped cross-section with a uniform thickness. Figure 8.26 shows SEM photographs of a single microgroove and groove arrays obtained using the two-step cutting method. The groove surfaces are extremely smooth and no burrs can be seen.

Figure 8.27 shows the SEM photographs in the front view and inclined view of micropyramid arrays fabricated on NiP surface with a 90° diamond tool. The height and width of the pyramids are 5 μm and 10 μm, respectively. Cutting was done by the two-step cutting method where the first cut was done at a depth of 4 μm and the second cut was done at a depth of 1 μm. No obvious burrs can be seen on the surface of the pyramids. Next, this kind of microstructured NiP surfaces were used as molding dies for glass molding tests.

Figure 8.26. *SEM photographs of microgrooves obtained by two-step cutting method: (a) a single groove, (b) groove arrays*

Figure 8.27. *SEM photographs in the front view and inclined view of micropyramid arrays fabricated on a NiP surface*

Glass molding experiments were conducted on an ultra-precision glass molding machine GMP211 produced by Toshiba Machine Co. Ltd., Japan. A flat NiP mold with machined microstructures and another flat NiP mold without microstructures were used as the lower and upper molds, respectively. Nitrogen gas was used to purge the air to prevent the molds from oxidation at high temperatures. The molding chamber was covered by a transparent silica glass tube which can let in the infrared rays from the heater and separate the nitrogen gas from the air outside. Temperatures of the upper and lower molds are monitored by two thermocouples beneath their surfaces. During molding, the upper mold remains stationary, and the lower mold is driven upward and downward by an AC servomotor. A load cell is placed beneath the lower axis as a feedback of the pressing load. A commonly used glass K-PG375 (Sumita Optical Glass Inc., Japan) of cylindrical shape was used as test piece material. The transition point (Tg) of the glass is 343°C and the yielding point (At) is 363°C. The molding temperature was set to 387°C.

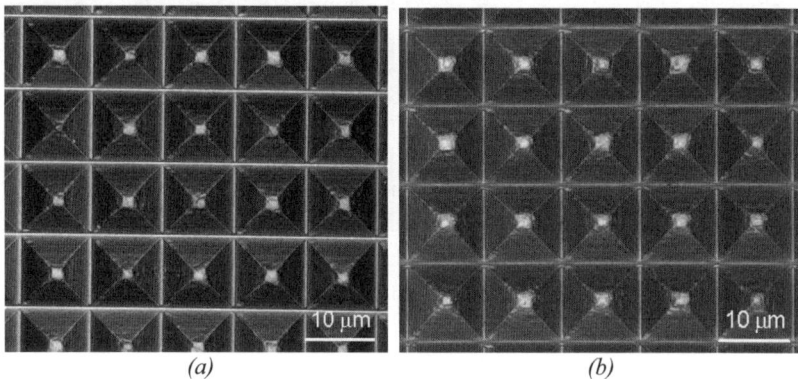

Figure 8.28. *Microscope images of micropyramid arrays on (a) NiP mold and (b) molded glass plate*

Figure 8.28 is a microscope image of micro-pyramid arrays on NiP mold surface and the pyramid arrays molded onto the glass surface, respectively. It is evident that the shape of the micro-pyramids on the NiP surface has been precisely transcribed to the glass surface. The surface of glass pyramids is as smooth as that of the NiP mold. The micro-pyramids may be used as new optical elements [LIN 98, LEE 06]. After molding micro-pyramid arrays for 50 shots, the NiP mold was taken out from the molding machine and observed using a microscope. No obvious damage and deformation of the mold was seen. It is presumably the case that the mold can be used for nearly 1,000 shots, similar to that in molding glass lenses.

8.4. Semiconductor and electronics related applications

Micro/nanomachining is also one of the important supporting technologies for semiconductor and electronics manufacturing industry. This involves not only the production technology of semiconductor wafers for fabrication of micro electronic parts, but also the planarization of the assembled LSI and ULSI substrates. In this section, we will briefly discuss these issues.

8.4.1. *Semiconductor wafer production*

Typical commercially important semiconductor materials include silicon (Si), germanium (Ge), gallium arsenic (GaAs), indium phosphide (InP) and silicon carbide (SiC). Most of these semiconductors are brittle crystalline materials, having very low fracture toughness; thus, they tend to behave in brittle manners during machining at room temperature. Currently, these materials are machined by abrasive machining processes such as grinding and lapping followed by chemo-mechanical polishing (CMP). Abrasive machining involves multi-point irregular mechanical contacts and simultaneous multi-scale material removal, and thus in turn is complicated and relatively difficult to control.

An alternative approach of wafer fabrication might be single-point machining processes represented by diamond turning. Diamond turning has been demonstrated to be capable of fabricating extremely smooth surfaces on brittle materials by controlling the microscopic material removal mode to be a completely ductile one, as has been shown in section 8.3. Compared to abrasive machining processes, diamond turning process has very high controllability.

However, presently there are two main problems which act as drawbacks for using diamond turning for fabricating silicon wafers. One is the problem of tool wear. Tool wear during machining silicon is very fast and involves very complex mechanisms of thermal, chemical, mechanical and possible electrical aspects, which is still under investigation [YAN 03]. Using suitable coolants is important to suppress the tool wear, but coolants are found to lower the ductile machinability. The development of new tool materials and new types of coolants would be the next subject in this area [OHT 08].

Another point is that ductile machining is not a damage-free machining process. Recent studies have revealed that even when machining is performed in a ductile mode, considerable subsurface damages will occur to the substrate material. The damages include microstructure changes, dislocations and residual stresses. Microlaser Raman [YAN 04-5, YAN 08-2] and TEM studies [YAN 09] of diamond-turned silicon wafers have revealed that amorphous layers will be generated beneath the machined surface. The subsurface damage caused by diamond turning is less than that caused by grinding but more severe than that generated by CMP processes. The elimination of subsurface damage in ductile machining of silicon is another important issue to study in this field.

Despite these problems, the feasibility of using diamond turning technology to other semiconductor wafer fabrication is promising. One of these semiconductors is single crystalline germanium (Ge). Germanium is as brittle as silicon, but the hardness is a little lower. Some encouraging results have been shown that in germanium machining, the tool wear is noticeably lower than that of silicon cutting [OHT 07]. This fact might be due to the chemical effects between germanium and diamond is considerably less significant than that between silicon and diamond. As a result, the tool life for ductile machining is very long and the production cost can be comparable to that of abrasive machining processes.

Ductile machining has also been used to fabricate compound semiconductor substrates, such as GaAs and InP, to substitute traditional grinding processes [YAN 06-2]. The compound semiconductors are generally softer than silicon, thus causing less wear to the diamond tools. It has been confirmed that surfaces with nanometer level surface roughness can be directly generated at high production efficiency without subsequent machining processes.

8.4.2. *LSI substrate planarization*

Finally, we will show an example application of diamond turning technology to LSI substrate planarization in electronics manufacturing industry. Recent LSI high-integration packaging technology, also called chip-on-chip (CoC) technology, requires the LSI chips to be 3D stacked into a few layers. The communication between these layers is accomplished by the micro-electrodes among these LSI chips.

The microelectrodes are usually made of soft metals such as gold (Au) or copper (Cu) by electro- or electroless-plating these materials onto large silicon wafers. After plating, the microelectrodes on the silicon wafer must be flattened to a very high flatness and low surface roughness, so that the pressing force required in chip joining can be reduced to, thereby protecting very low in order the electronic circuits from damages or fractures. The flattening technology of the micro-electrodes, also called planarization technology, is essential in CoC manufacturing industry. Conventionally, the planarization has been done by CMP processes. However, there are many problems such as low production efficiency, low flatness and environmental pollutions.

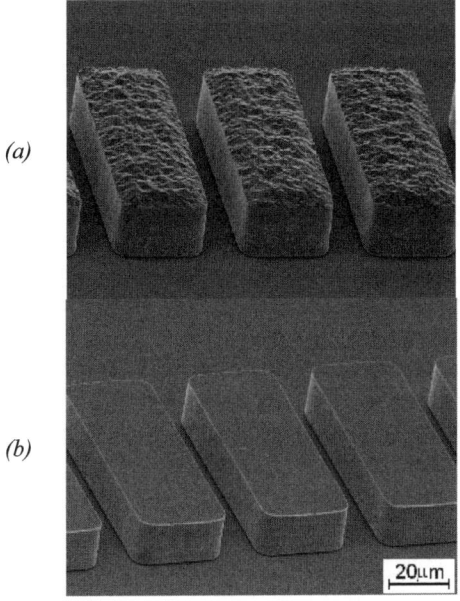

Figure 8.29. *SEM micrographs of the micro-electrodes (a) before and (2) after planarization*

Diamond turning has been successfully used to perform planarization of LSI substrates to solve the above-mentioned problems [YAN 05-2]. Of course, this technology requires renovation in the design structure of diamond turning machines and tool geometry to achieve the most cost-efficient performance. The details to realize this technology will not be given in this book but may be found in other publications of the present author. As an example of the results possible, Figure 8.29 shows two SEM micrographs of the micro-electrodes before and after planarization. It is clear that by diamond turning, the surfaces of microelectrodes have been flattened to a nanometer level surface roughness and extremely high flatness. This method is very easy to control automatically and the production efficiency is far higher than the CMP process.

8.5. Summary

Micro/nanomachining has become a subject of concentrated research interest in the optical and optoelectronics manufacturing industries. In this chapter, some examples of micro/nanocutting technologies for metals, semiconductors and crystalline materials have been introduced and discussed based on the recent research outcomes of the author's research group in collaboration with industry. It should be pointed out again that micro/nanomachining is not confined to nano-cutting or diamond turning. A large variety of other high-precision machining technologies are serving the modern high-tech manufacturing industry. The research and development in this area will be endless.

8.6. Acknowledgements

The author would like to extend his gratitude to his colleagues and graduate students in his research group and industrial partners for their support and help in the research projects. The studies presented in this chapter have been carried out with the continuing financial support from Japanese industry, Japan New Energy and Industrial Technology Development Organization (NEDO), the Japan Ministry of Education, Culture, Sports, Science and Technology (MEXT), and the Japan Science and Technology Agency (JST).

8.7. References

[BLA 90] BLAKE P.N., SCATTERGOOD R. O., "Ductile regime machining of germanium and silicon", *J. Amer. Ceram. Soc.*, Vol. 73, No. 4 (1990) p. 949-957.

[BRY 67] BRYAN J. B., CLOUSER R. R., HOLLAND E, "Spindle accuracy", *Amer. Mach. Spec. Rep.*, No. 612, (1967) p. 149.

[CHA 94] CHANG R., CHERN T., LIN C., LAY Y., "Fabrication and testing of high quality small germanium plano-convex lens", *Opt. Laser. Eng.*, Vol. 21, No. 5 (1994) p. 257-272.

[IKA 91] IKAWA N., DONALDSON R. R., KOMANDURI R., KONIG W., MCKEOWN P. A., MORIWAKI T., STOWERS I. F., Ultra-precision metal cutting – the past, the present and the future", *Ann. CIRP*, Vol. 40, No. 2 (1991) p. 587.

[JOH 05] JOHNSON R.B., MICHAEL M., "Aspheric glass lens modeling and machining", in *Proc. of SPIE, Current Developments in Lens Design and Optical Engineering VI*, San Diego, CA, USA, 58740B-58740Q, 2005.

[KAP 02] KAPLAN S. G., HANSSEN L. M., "Silicon as a standard material for infrared reflectance and transmittance from 2 to 5 μm", *Infra. Phys. Tech.*, Vol. 43 (2002) p. 389-396.

[KAT 06] KATSUKI M., "Transferability of glass lens molding", in *Proc. of SPIE, 2nd International Symposium on Advanced Optical Manufacturing and Testing Technologies*, Xi'an, China, 61490M-61490V, 2006.

[KRA 69] KRAAKMAN H. J. J., GAST J. G. C., "A precision lathe with hydrostatic bearings and drive", *Philips Tech. Rev.*, Vol. 30, No. 5, (1969) p. 117.

[LEE 06] LEE J. Y., KIM Y. J., NAHM K. B., KO J. H., "Optical simulation of micro-pyramid arrays for the applications in the field of backlight unit of LCD", in *IMID/IDMC '06 DIGEST*, (2006) p. 1343-1346.

[LEU 98] LEUNG T.P., LEE W.B., LU X.M., "Diamond turning of silicon substrates in ductile-regime", *J. Mater. Proc. Tech.*, Vol.73, (1998) p. 42.

[LIN 98] LIN L., CHENG Y.T., CHIU C. J., "Comparative study of hot embossed micro structures fabricated by laboratory and commercial environments", *Micro System Tech.*, Vol. 4 (1998) p.113-116.

[MAS 07] MASUDA J., YAN J., KURIYAGAWA T., "Application of the NiP-plated steel molds to glass lens molding", *Proc. of ISAAT, Advances in Abrasive Technology*, Dearborn, USA, (2007) p.123-130.

[MOR 95] MORRIS J. C., CALLAHAN D. L., KULIK J., PATTEN J. A., SCATTERGOOD R. O., "Origins of the ductile regime in single-point diamond turning of semiconductors", *J. Am. Ceram. Soc.*, Vol. 78, No.8 (1995) p.2015-2020.

[NAK 90] NAKASUJI T., KODERA S., HARA S., MATSUNAGA H., IKAWA N., SHIMADA S., "Diamond turning of brittle materials for optical components", *Ann. CIRP*, Vol. 39, No.1 (1990) p. 89-92

[NIC 81] NICHOLAS D.J., BOON J.E., "The generation of high precision aspheric surfaces in glass by CNC machining", *J. of Physics D, Applied Physics*, Vol. 4, (1981) p. 593-600.

[OHT 07] OHTA T., YAN J., YAJIMA S., TAKAHASHI Y., HORIKAWA N., KURIYAGAWA T., "High-efficiency machining of single-crystal germanium using large-radius diamond tools", *Int. J. Surf. Sci. Eng.*, Vol.1, No. 4, (2007) p. 374–392.

[OHT 08] OHTA T., YAN J., YAJIMA S., TAKAHASHI Y., HORIKAWA N., KURIYAGAWA T., "Coolant effects on tool wear in machining single-crystal silicon with diamond tools", *Key Eng. Mater.* (in press)

[SHE 85] SHEA D.C.O., *Elements of Modern Optical Design*, Wiley, New York, 1985.

[SYN 98] SYN C. K., "An empirical survey on the influence of machining parameters in diamond turning of large single crystal silicon optics", in *Proc. ASPE Spring Topical Meeting on Silicon Machining*, April (1998) p. 44.

[TEE 99] TEEGARDEN B. J., "Space instrumentation for gamma-ray astronomy", *Nucl. Instrum. Meth. Phys. Res.* A, Vol. 422, No. 1-3 (1999) p. 551-561.

[WHI 66] WHITTEN L.G. LEWIS T.G., "Machining and measurement to submicron tolerance", *Proc. M.T.D.R.* (1966) p. 491.

[YAN 98] YAN J., SYOJI K., SUZUKI H., KURIYAGAWA T., "Ductile regime turning of single crystal silicon with a straight-nosed diamond tool", *J. Jap. Soc. Prec. Eng.*, Vol. 64, No. 9 (1998) p. 1345.

[YAN 01] YAN J., YOSHINO M., KURIYAGAWA T., SHIRAKASHI T., SYOJI K., KOMANDURI R., "On the ductile machining of silicon for micro electro-mechanical Systems (MEMS), opto-electronic and optical applications", *Mater. Sci. Eng. A*, Vol. 297, No .1-2 (2001) p. 230.

[YAN 02-1] YAN J., SYOJI K., KURIYAGAWA T., SUZUKI H., "Ductile regime turning at large tool feed", *J. Mater. Proc. Tech.*, Vol. 121, No. 2-3 (2002), p.363.

[YAN 02-2] YAN J., SYOJI K., KURIYAGAWA T., "Fabrication of large-diameter single-crystal silicon aspheric lens by the straight-line enveloping diamond-turning method", *J. Jpn. Soc. Prec. Eng.*, Vol. 68, No. 4, (2002) p. 1561-565.

[YAN 03] YAN J., SYOJI K., TAMAKI J., "Some observations on the wear of diamond tools in ultra-precision cutting of single-crystal silicon", *Wear*, Vol. 255, No. 7-12 (2003) p. 1380-1387.

[YAN 04-1] YAN J., SASAKI T., TAMAKI J., KUBO A., SUGINO T., "Chip formation behaviour in ultra-precision cutting of electroless nickel plated mold substrates", *Key Eng. Mater.*, Vol. 257-258 (2004) p. 3-8.

[YAN 04-2] YAN J., MAEKAWA K., TAMAKI J., KUBO A., "Experimental study on the ultraprecision ductile machinability of single-crystal germanium", *JSME Int. J.*, C, Vol. 47, No. 1 (2004) p. 29-36.

[YAN 04-3] YAN J., SYOJI K., TAMAKI, J, "Crystallographic effects in micro/nanomachining of single-crystal calcium fluoride", *J. Vacuum Sci. Tech. B*, Vol. 22, No. 1 (2004) p. 46-51.

[YAN 04-4] YAN J., SYOJI K., TAMAKI, J, KURIYAGAWA T., "Single-point diamond turning of CaF_2 for nanometric surfaces", *Int. J. Adv. Manuf. Techn.*, Vol. 24, No. 9-10 (2004) p. 640-646.

[YAN 04-5] YAN J., "Laser micro-Raman spectroscopy of single-point diamond machined silicon substrates", *J. App. Phys.*, Vol. 95, No. 4 (2004) p. 2094–2101.

[YAN 05-1] YAN J., MAEKAWA K., TAMAKI J., KURIYAGAWA T., "Micro grooving on single-crystal germanium for infrared Fresnel lenses", *J. Micromech. Microeng.*, Vol. 15 (2005) p. 1925–1931.

[YAN 05-2] YAN J., "Recent advances in ultraprecision cutting technology", *Proc. of the 2005 JSPE Spring Meeting* (2005) p. 761-762.

[YAN 06-1] YAN J., TAKAHASHI Y., TAMAKI J., KUBO A., KURIYAGAWA T., SATO Y., "Ultraprecision machining characteristics of poly-crystalline germanium", *JSME Int. J.*, C, Vol. 49, No. 1 (2006) p. 63-69.

[YAN 06-2] YAN J., ZHAO H., KURIYAGAWA T., TAMAKI J., "Nanoindentation and diamond turning tests on compound semiconductor InP", in *Proc. 6th Euspen Int. Conf.*, Baden bei Wien, Austria, May (2006) p. 276-279.

[YAN 08-1] YAN J., OOWADA T, ZHOU T., KURIYAGAWA T., Precision machining of microstructures on electroless-plated nip surface for molding glass components, in *Proc. 8th Asia-Pacific Conference on Materials Processing,* 2008 (CD-ROM).

[YAN 08-2] YAN J., ASAMI T., KURIYAGAWA T., Nondestructive measurement of the machining-induced amorphous layers in single-crystal silicon by laser micro-Raman spectroscopy, *Prec. Eng.*, 32 (2008) p. 186-195.

[YAN 09] YAN J., ASAMI T., KURIYAGAWA T., Fundamental investigation on subsurface damages in single crystalline silicon caused by diamond machining, *Prec. Eng.* (2009) (submitted).

[YU 94] YU J., YAN J., MA W., HAN R., "Ultra-precision diamond turning of optical crystals", in *Proceedings of the SPIE,* Vol. 1994 (1993) p. 51-59.

List of Authors

Waqar AHMED
School of Chemistry, Computing and Technology
University of Central Lancashire
Preston
UK

J. Paulo DAVIM
Department of Mechanical Engineering
University of Aveiro
Portugal

Wit GRZESIK
Department of Manufacturing Engineering and Production Automation
Technical University of Opole
Poland

Mark J. JACKSON
Center for Advanced Manufacturing
College of Technology
Purdue University
West Lafayette, Indiana
USA

Xiaoping LI
Department of Mechanical Engineering
National University of Singapore
Singapore

Zhijian PEI
Department of Industrial and Manufacturing Systems Engineering
Kansas State University
Manhattan, Kansas
USA

Rüdiger RENTSCH
Department of Production Engineering
Bremen University
Germany

Grant M. ROBINSON
Center for Advanced Manufacturing
College of Technology
Purdue University
West Lafayette, Indiana
USA

Jiangang SUN
Nuclear Engineering Division
Argonne National Laboratory
Argonne, Illinois
USA

Michael D. WHITFIELD
Micro Machinists
LLC
West Lafayette, Indiana
USA

Jiwang YAN
Department of Nanomechanics
Tohoku University
Sendai
Japan

Jianmei ZHANG
Department of Industrial Engineering
University of Texas at El Paso
El Paso, Texas
USA

Index

3D MD cutting model, 10

A-C

atomic interaction, 4-6
burr formation, 92
chemical vapor deposition (CVD), 45-52, 55, 58, 59, 61, 62, 66, 67
crack initiation, 35
cutting forces, 13-16

D

diamond, 45-67
diamond turning, 74, 176
ductile machining, 181

F-I

femtosecond, 128, 131, 132, 133, 143, 146, 147, 148, 149, 150, 154
grinding, 101, 104, 105, 106, 107, 108, 109, 110, 111, 120, 121, 122
hydrostatic pressure, 34
inspection technique, 92

L

laser, 104, 107, 109, 125, 128, 130, 131, 133, 136, 144, 145, 146, 147, 150, 151, 154, 155

M

machined surfaces, 38
machining, 129, 134, 135, 136, 138, 139, 140, 143, 146, 147, 148, 154
machining accuracy, 72
machining center, 90
manufacturing, 101
material, 101, 102, 103, 104, 105, 106, 107, 108, 109, 113, 115, 121
MD simulation, 8, 11, 18, 19
melt, 131, 132, 133, 134, 138, 150
MEMS, 157
micro/nanomachining, 175
microcutting, 71, 79
microcutting tools, 64
microfabrication, 133, 138, 143, 146
microgrinding, 101
micromachining, 71, 125, 129, 137, 143, 150, 157
micromilling, 85, 87
microscale, 104, 125, 154
microstructures, 193

miniature tool, 75
molecular dynamics (MD), 1-13, 15-23

N

nanofabrication, 151, 154, 155
nanogrinding, 101, 104, 105, 106, 107, 121
nanomanufacturing, 151, 154
nanoscale, 104, 121
nanoscale cutting, 3, 4, 12
nanosecond, 131, 133, 134, 138, 148
nanostructures, 153
non-destructive evaluation, 157

O-P

optical elements, 179
phase transformation, 34
physical vapor deposition (PVD), 57, 67
picosecond, 143, 144, 146, 147

plasma, 132, 135
polishing, 101, 102, 105
precision technology, 71

S

semiconductor, 200
spot size, 129, 136
stress and temperature, 15-18
stress conditions, 36
subsurface damage, 39, 157
surface damage, 157
surface roughness, 75
Swiss lathe, 73
system dynamics, 7-8

T-W

tools, 101, 104, 106, 121
tungsten carbide, 50, 51, 59
ultra-precision, 101, 102, 103
vitrified, 106, 108, 109, 111, 120, 121
wear, 101, 104, 106, 108, 120, 121